ANOMALY

ANOMALY
A Scientific Exploration of the UFO Phenomenon

Daniel Coumbe

ROWMAN & LITTLEFIELD
Lanham • Boulder • New York • London

Published by Rowman & Littlefield
An imprint of The Rowman & Littlefield Publishing Group, Inc.
4501 Forbes Boulevard, Suite 200, Lanham, Maryland 20706
www.rowman.com

86-90 Paul Street, London EC2A 4NE

Copyright © 2023 by The Rowman & Littlefield Publishing Group, Inc.

All rights reserved. No part of this book may be reproduced in any form or by any electronic or mechanical means, including information storage and retrieval systems, without written permission from the publisher, except by a reviewer who may quote passages in a review.

British Library Cataloguing in Publication Information Available

Library of Congress Cataloging-in-Publication Data

Names: Coumbe, Daniel, author.
Title: Anomaly : a scientific exploration of the UFO phenomenon / Daniel Coumbe.
Description: Lanham : Rowman & Littlefield, [2023] | Includes bibliographical references and index. | Summary: "Anomaly: A Scientific Exploration of the UFO Phenomenon reveals new results derived from radar, optical sensors, and scientific instruments, rather than speculating on unreliable eyewitness testimony. This scientific approach provides the reader with clear and reliable answers, something that is desperately needed in the murky field of UFOs"—Provided by publisher.
Identifiers: LCCN 2022031578 (print) | LCCN 2022031579 (ebook) | ISBN 9781538172131 (cloth) | ISBN 9781538172155 (ebook)
Subjects: LCSH: Unidentified flying objects.
Classification: LCC TL789 .C677 2022 (print) | LCC TL789 (ebook) | DDC 001.942—dc23/eng/20220902
LC record available at https://lccn.loc.gov/2022031578
LC ebook record available at https://lccn.loc.gov/2022031579

∞™ The paper used in this publication meets the minimum requirements of American National Standard for Information Sciences—Permanence of Paper for Printed Library Materials, ANSI/NISO Z39.48-1992.

To Ivana, Ava, and Noa,
for forever changing the way I see the world.

Contents

Preface		xi
1	We Need to Talk about UFOs	1
	1.1 Black Swans	1
	1.2 There's Something in the Air	3
	1.3 Filtering Evidence	8
	1.3.1 Filter 1: Eyewitness Testimony	8
	1.3.2 Filter 2: Single Sensor Data	10
	1.3.3 Filter 3: Multiple Sensor Data	10
	1.3.4 Filter 4: Physical Evidence	11
	1.3.5 Summary	11
	1.4 Notes	12

Part I: Case Studies

2	Japan Airlines Flight 1628	17
	2.1 Case Description	17
	2.2 Analysis	25
	2.2.1 How to Track a UFO on Radar	25
	2.2.2 Remarkable Results	26
	2.2.3 Evaluation of Evidence	31
	2.3 Appendix	34
	2.3.1 Kinematics	34
	2.3.2 Processing the Radar Data	35

		2.3.3 Error Estimates	36
		2.3.4 Data	37
	2.4	Notes	37
3	The Brazilian Fragments		39
	3.1	Case Description	39
	3.2	Analysis	42
		3.2.1 Bulk Composition	42
		3.2.2 Impurities	45
		3.2.3 Isotopes	47
		3.2.4 Evaluation of Evidence	49
	3.3	Appendix	52
		3.3.1 Uncertainty of Terrestrial Isotopic Ratios	52
		3.3.2 Uncertainty of Cleveland 2018 Isotopic Ratios	53
		3.3.3 Sigma Tension	53
	3.4	Notes	54
4	The Lonnie Zamora Incident		57
	4.1	Case Description	57
	4.2	Analysis	61
		4.2.1 Weighing a UFO	61
		4.2.2 Testing Alternative Hypotheses	62
		4.2.3 Weighing an Alien	67
		4.2.4 Evaluation of Evidence	68
	4.3	Appendix	70
		4.3.1 How to Weigh a UFO	70
		4.3.2 Resolution of the Human Eye	72
		4.3.3 The Physics of Balloons	72
	4.4	Notes	73
5	The Aguadilla Object		75
	5.1	Case Description	75
	5.2	Analysis	76
		5.2.1 Radar Data	76
		5.2.2 Thermal Imaging Data	78
		5.2.3 Evaluation of Evidence	82
	5.3	Appendix	84
		5.3.1 Haversine Formula	84
		5.3.2 Error Estimates	84
	5.4	Notes	85

Part II: The Bigger Picture

6	Where Are All the UFOs?		89
	6.1	A UFO Toolkit	89

	6.2	A UFO Database		98
	6.3	Military		101
		6.3.1	Introduction	101
		6.3.2	Analysis	104
	6.4	Water		108
		6.4.1	Introduction	108
		6.4.2	Analysis	110
	6.5	Environment		115
		6.5.1	Introduction	115
		6.5.2	Analysis	116
	6.6	Earthquakes		119
		6.6.1	Introduction	119
		6.6.2	Analysis	122
	6.7	Blood		124
		6.7.1	Introduction	124
		6.7.2	Analysis	125
	6.8	Appendix		126
		6.8.1	Statistical Calculations	126
		6.8.2	Military Study	128
		6.8.3	Water Study	128
		6.8.4	Environment Study	129
		6.8.5	Earthquake Study	130
		6.8.6	Blood Study	130
	6.9	Notes		132
7	When Are All the UFOs?			135
	7.1	Introduction		135
	7.2	Analysis		140
	7.3	Appendix		150
		7.3.1	Calculations	150
		7.3.2	E.T. Movies Study	150
		7.3.3	Time of Year Study	151
	7.4	Notes		151
8	What Does It All Mean?			153
	8.1	What Are They?		153
	8.2	What Do They Want?		158
	8.3	What Now?		160
	8.4	Notes		162

Index 165

About the Author 169

Preface

THE PURPOSE OF *Anomaly: A Scientific Exploration of the UFO Phenomenon* is to provide one of the first legitimate scientific explorations of the UFO phenomenon in the modern era. Considering the recent Pentagon admissions on UFOs, such a scientific approach is desperately needed in a field shrouded in hearsay, anecdotes, and speculation. The UFO phenomenon is an enigma wrapped in a mystery, and our best chance of unraveling this Gordian Knot is science. I hope this book will also encourage other scientists to take up the mantle.

This book is for anyone that wants real answers to the UFO phenomenon supported by hard scientific evidence. You may have to think a little more than in most books on this subject, but you will be rewarded with more reliable insights. This book is aimed at the intelligent layman. The only requirements are a child-like curiosity, an ability to ask difficult questions, and the courage to accept the answer even if it does not fit your preconceived ideas.

This book is divided into two parts. Part I involves the analysis of four of the best-evidenced and well-documented UFO incidents. Each of these four case studies is presented in an approachable and non-technical manner. All technical details have been moved to an appendix at the end of each chapter. In two of these case studies, we analyze real radar data from UFO encounters, and obtain startling new results that suggest UFOs exhibit extraordinary flight characteristics that exclude the possibility of piloted craft. In another case, we analyze the chemical composition of metallic fragments that allegedly came from a UFO. These fragments are found to have a highly unusual chemical composition, especially given the context in which they were found.

In the last case, we analyze indentations left by the purported landing of a UFO, which allows us to estimate the weight of a UFO for the first time. Possible footprints were also discovered in this case, allowing us to weigh whatever left them.

In Part II of this book, possible motives behind the UFO phenomenon are explored. We investigate the potential connection between the number of UFO sightings and: the military, water, the environment, earthquakes, blood type, and movie releases featuring extraterrestrials. We uncover two statistically significant correlations, one of which has never been revealed before. The timing of UFO sightings is also investigated. An interesting result from this analysis includes a statistically significant spike in UFO reports on the infamous date of September 11, 2001.

The idea for this book was conceived in December 2017, when I first heard the extraordinary eyewitness testimony of David Fravor, the fighter pilot who claimed visual contact with a UFO during the 2004 *USS Nimitz* incident. His pedigree and sincerity made me stop and think, *could there be something to this*? For the next three years, I plunged into the UFO literature, with a laser-like focus on data rather than personal testimony. I wanted to remove human error from the equation and settle this matter once and for all. *Data doesn't lie*, was my mantra. This book is the result.

Given my background, I was confident that I could contribute to the scientific study of this phenomenon. I have a PhD in theoretical particle physics and six years' experience as a research fellow in the field of quantum gravity, including four years at the prestigious Niels Bohr Institute in Copenhagen. I use this knowledge and experience to uncover new insights and results that may help to finally unravel the UFO enigma.

1
We Need to Talk about UFOs

What is true, and I'm actually being serious here, is that there are, there's footage and records of objects in the skies, that we don't know exactly what they are. We can't explain how they moved, their trajectory. They did not have an easily explainable pattern.

—Barack Obama (forty-fourth president of the United States)[1]

1.1 Black Swans

ON JANUARY 10, 1697, a Dutch explorer and sea captain called Willem de Vlamingh saw something unthinkable. Captain Vlamingh and his crew were exploring the coast of western Australia by ship and, on a whim, decided to sail up one of the rivers. Out on deck, Vlamingh happened to glance down into the river below. A single black swan starred up at him. To the black swan, this moment was like any other. To Vlamingh, it was a revelation.

You see, before this moment, everybody from the old world *knew* that all swans were white. It was an obvious fact confirmed by everyday experience. Yet, there sat a black swan. It was undeniable. Suddenly, Vlamingh understood. His limited experience as a European led him to assume that all swans are white, because that's all he had ever seen. This model was now broken, forever. There was no choice in the matter, he must replace what he thought he knew with a better model, a broader perspective.

This is not about swans, but something far bigger. It illustrates how we come to know things about the world around us through the scientific method. The point is that a model of our world built on centuries of empirical data can be shattered by a single observation. All it takes is one black swan.

In science, black swans are referred to as anomalies; results that buck the trend defined by the majority of data. They are the outlier that refuses to conform. Or, more formally, an anomaly is a statistically significant difference between a theoretical prediction and an observed result. Anomalies have played, and continue to play, a decisive role in the structure of scientific revolutions. To quote Thomas Kuhn:

> Discovery begins with the awareness of anomaly—the recognition that nature has violated the paradigm-induced expectations that govern normal science. The area of the anomaly is then explored. The paradigm change is complete when the paradigm has been adjusted so that the anomalous become the expected.[2]

For example, in 1859, a small problem was found with the orbit of the planet Mercury. All other planets in the solar system dutifully obeyed Newton's laws of physics. Like clockwork, they would predictably and obediently orbit the Sun. But Mercury stubbornly defied Newton. Urbain Le Verrier, a French astronomer and mathematician, showed that a specific feature of Mercury's orbit disagreed with Newtonian mechanics by around forty arc-seconds per century. A small discrepancy, granted, but one that refused to go away. At first, this result was like a harmless scratch in Newton's armor. But under the intense and sustained pressure of scientific scrutiny, it opened into a crack, a crack that would deepen and slowly spread until the entire theory was fractured beyond repair. This one bit of anomalous data contributed to the downfall of Newton's law of universal gravitation, which had reigned supreme for nearly 250 years, and ushered in a new paradigm: Einstein's general theory of relativity. This is the awesome power of a black swan result—it can spark a scientific revolution.

Most apparent anomalies, however, turn out to be mirages that fade under closer inspection. For example, data collected at the Large Hadron Collider (LHC) at CERN in 2015 displayed a strong anomaly. The LHC smashes protons into other protons at incredibly high energies and then scours the debris from these high-speed wrecks for signs of new particles. The analyzed data hinted at the existence of a new particle, one that could upend existing models of particle physics. Anomalies are quantified by their so-called sigma value. A higher sigma value means a higher probability that the anomaly is real, and not just a statistical fluke. The initial LHC result had a sigma value of about 3.5, which equates to roughly a one in five thousand chance that the anomaly was not real. Good odds. Good enough for the physics community to boil over with excitement, leading to an astonishing five hundred publications on this result in just nine months. However, as more data trickled in, this sigma value began to slowly fade until eventually, in 2016, enough data had accumulated to show that this initially anomalous result was nothing more than a statistical fluke. There was no revolution to be had. Science had to go on as normal.

A good scientist does not ignore anomalous data, they investigate it with renewed intensity, hoping to overthrow the existing paradigm and ignite a new revolution. Most of the time the apparent anomaly does not hold up and fades away into insignificance, but sometimes you get lucky. This is how science progresses. Profound leaps in our understanding of the world can be made during these extremely rare paradigm shifts. Realistically, scientists do not have the time, funding, or inclination to investigate every single piece of potentially anomalous data. However, what typically happens is that evidence slowly mounts until a kind of tipping point is reached, only then does a particular anomaly warrant a full scientific investigation. In the case of UFOs, we may be getting close to such a tipping point.

1.2 There's Something in the Air

December 17, 2017, the front page of the *New York Times* reads: "Real UFOs? Pentagon Unit Tried to Know." Why did the acclaimed newspaper, with more Pulitzer Prizes than any other, feel compelled to tackle a subject with so much stigma? The answer is simple. Because they uncovered something big, and they had proof.

In October 2017, Luis Elizondo resigned as the director of a shadowy Pentagon program. On the very day Elizondo resigned, he met with journalist Leslie Kean in a hotel lobby in Washington, DC. Kean would later describe the meeting as "life-changing."[3] For two hours, Elizondo revealed "stunning information," provided extensive supporting documentation, and even showed Kean three videos of UFOs captured by the US Navy. Armed with this material, Kean and a colleague, Ralph Blumenthal, pitched the idea of an article to the head editor of the *New York Times*. They got the green light.

The *Times* article exposed a $22 million black budget program run by the Pentagon to investigate UFOs. The covert program, known as the Advanced Aerospace Threat Identification Program (AATIP), investigated reports of UFOs between 2007 and 2012, contrary to previous denials by the government. But that was not all. Attached to the article were the videos Leslie Kean had seen during that life-changing meeting. The videos show encounters between US fighter jets and unidentified craft displaying highly unusual flight characteristics. Since the publication of this article, the US Department of Defense (DoD) has officially acknowledged the existence of the AATIP program and confirmed that the three videos are authentic. On August 4, 2020, Deputy Secretary of Defense David Norquist established the Unidentified Aerial Phenomena Task Force (UAPTF), a successor of the AATIP program, a covert program that continues to this day.

Why does the US government continue to spend tens of millions of dollars investigating UFOs, and why has it done so for well over a decade in complete secrecy? According to a statement by the DoD, "The mission of the UAPTF is to detect, analyze and catalog UAPs that could potentially pose a threat to US national security."[4] Is the government now taking UFOs seriously because of a possible threat to the US? Understanding the context in which the three Pentagon UFO videos were captured may shed light on this.

The first of the three videos were taken in November 2004 from a fighter jet based on the aircraft carrier USS *Nimitz*, the lead vessel of a naval strike group conducting pre-deployment exercises off the northwest coast of Mexico. Starting on November 10, the USS *Princeton*, the nerve center of the naval strike group, began tracking objects on the radar that displayed highly unconventional flight patterns. These anomalies continued to appear on radar for the next four days.[5] Thinking it was a fault, technicians rebooted the entire radar system and ran extensive diagnostic tests. The tracks remained. The strike group had an air defense exercise scheduled for November 14. Senior Chief Kevin Day, a radar operator aboard the USS *Princeton*, grew concerned and warned the captain that the UFOs may pose a flight safety risk during the exercise. Two F/A-18F Super Hornet fighter jets were re-routed from their duties to intercept one of the UFOs. Following visual contact, the UFO disappeared. As if taunting them, only five seconds later, radar picked it up at the fighter jet's secret rendezvous point, some sixty miles away. A second wave of jets was launched from the *Nimitz* to investigate. One of the fighter pilots in this second wave, Commander Chad Underwood, managed to capture the first of these Pentagon UFO videos using an advanced forward-looking infrared (FLIR) camera. The UFOs, whatever they were, clearly hindered the strike group's ability to carry out vital pre-deployment exercises. Not to mention the fact that they somehow knew the encrypted rendezvous point.

The potential threat to national security became even more apparent during a series of encounters off the southeast coast of the US between the summer of 2014 and March 2015. During this time, radar and visual encounters were reported on an "almost daily basis" by members of the naval strike group led by the USS *Theodore Roosevelt*.[6] In late 2014, two Super Hornet fighter pilots were on an exercise flying just one hundred feet apart when a UFO shot rapidly between them. The pilots, shaken by the near mid-air collision, immediately reported the incident. An official safety report was also placed by the squadron. It was during these frequent and alarming series of encounters that the second and third Pentagon videos were captured.

Possibly in response to these incidents, the Navy declared that it is "updating and formalizing the process" of reporting UFO encounters in the spring of 2019. In June 2021, the Office of the Director of National Intelligence (ODNI) called for yet further measures, including a consolidation of reports from across the federal government, increased collection and analysis of UFO data, and additional funding for research and development. At least within the military, the once ridiculed and taboo subject of UFOs is now clearly a serious topic of concern.

A significant shift has also taken place among US political leadership. For example, it turns out that the 22-million-dollar black budget for the AATIP program came at the request of Senate Majority Leader Harry Reid, with additional backing from senators Ted Stevens and Daniel Inouye. Throughout 2019, numerous senators received classified Pentagon briefings about UFO encounters, including Senator Mark Warner, the vice-chairman of the Senate Intelligence Committee.[7] In June 2019, then President Donald Trump revealed that he had also been briefed on UFOs. It was not long before these briefings led to action. Hidden away in an obscure comment section of the huge 2.3 trillion-dollar coronavirus relief bill signed in December 2020, was a demand from senator and presidential candidate Marco Rubio, a demand that US intelligence agencies disclose everything they know about UFOs within 180 days. A further stipulation was that the report must be unclassified and made available to the public.

The highly anticipated report was finally released on June 25, 2021. It was both disappointing and revelatory. Disappointing because it was just nine pages long and did not tell us what UFOs are. Revelatory because it told us what UFOs are not. The report definitively ruled out the possibility that UFOs are secret US military technology, a possibility long considered by many as the most likely explanation. The document deals with a total of 144 UFO reports originating from US military and intelligence sources, eighty of which were captured by multiple sensors. In twenty-one of these reports, unusual movement patterns and/or flight characteristics were observed and recorded.[8] Alarmingly, eleven of these instances involved near mid-air collisions with a UFO. In 143 of the 144 cases studied, the UFO could not be identified.

So, according to this Pentagon report, between 2004 and 2021 there were eighty UFO incidents registered across multiple sensor platforms. This equates to an average of about five incidents every year for seventeen consecutive years. And need I remind you that these sensors are among the most advanced in the world, including state-of-the-art radar systems, infrared cameras, electro-optical devices, and weapon seekers.[9] If this abundance of data were readily available it could, at the very least, definitively demonstrate that UFOs are a statistically significant and persistent anomaly. Frustratingly, although the Pentagon states that this data exists, it remains classified. Yet, there is a smaller amount of unclassified data available on UFOs. Even with this restricted data set, it may still be possible to determine whether UFOs constitute a true anomaly.

I do not know what UFOs are, but I do know that it is not the job of the military or politicians to find out—it is the job of the scientist. Science is the most powerful tool we have for understanding the world, and we must bring all its might to bear on this intriguing phenomenon. The primary aim of this book is to present the reader with the best unclassified evidence available, in the form of raw and processed data. You, the reader, can then come to a better-informed conclusion as to whether UFOs constitute a real anomaly. The secondary aim of this book is to help initiate a credible scientific study of the UFO phenomenon. This task is too big for one book. I hope that this work will inspire other trained scientists to conduct serious data-driven scientific investigations into the UFO phenomenon.

How did I first get interested in this topic? It wasn't one event but, rather, a gradual accumulation of factors. One pivotal moment was when I first heard the extraordinary eyewitness testimony of Commander David Fravor, the fighter pilot who claims he had visual contact with a UFO during the 2004 USS *Nimitz* incident. His seemingly sincere recollection of events made me stop and think, *could there be something to this?* Commander Fravor reported encountering a forty-foot-long craft, shaped like a smooth white tic-tac, making highly erratic movements just above the surface of the Pacific Ocean.[10] This man is a graduate of the Top Gun Academy with thousands of hours of experience flying state of the art fighter jets in high-pressure combat scenarios. A man we trust to protect our lives. Not a man one would expect to be prone to fantasy. Plus, he had a lot to lose in coming out with such a story, including his reputation, and apparently not much to gain. Until that point, I was a die-hard sceptic of UFOs. But there was something about his sincerity and pedigree that made me think twice.

Being a scientist, I then doubted myself—for three whole years. During this time, I plunged into the UFO literature, with a laser focus on data rather than people's testimonies. I wanted to remove human error from the equation and settle this matter once and for all. As convincing as the testimony from this highly trained fighter pilot was, I wanted to be sure, and data doesn't lie. This book is the result. To be candid, it is a purely selfish attempt to find out whether there is anything to this topic.

Given my background, I was confident that I could contribute something to the scientific study of this phenomenon. I have a PhD in theoretical particle physics from the University of Glasgow in Scotland. Following my PhD, I spent a further six years as a research fellow in the field of quantum gravity, including four years at the prestigious Niels Bohr Institute in Copenhagen. I aim to use this knowledge and experience to estimate the evidence for, and physical capabilities of, UFOs. Crucially, this scientific background has schooled me in the importance of remaining impartial, removing sources of bias, carefully quantifying measurement errors, and letting the evidence speak for itself. I hope to provide one of the first credible data-driven scientific studies of UFOs in the modern era.

This book is aimed at anyone serious about getting to the bottom of the UFO phenomenon. The best tool we have for doing this is science. You may be required to think a little more analytically than in most other books on this subject, but you will be rewarded with more reliable insights. Each chapter in the first part of this book focuses on a different UFO incident and is generally divided into two sections: a case description and a case analysis. Only the main results and conclusions of the study are presented in the analysis section, with all technical details relegated to an appendix at the end of each chapter for the interested reader. The second part of the book takes a much broader perspective, looking for possible relationships between UFOs and several popular hypotheses. The relevant concepts will be built up slowly and intuitively. The only real requirements are a child-like curiosity, an ability to ask difficult questions, and the courage to accept the answer, whether it fits with your preconceived ideas or not.

Assuming UFOs are real physical objects, not optical illusions or weather phenomena, then there are essentially three possibilities. Either UFOs are (1) secret US technology, (2) secret non-US technology, or (3) non-man-made technology. The problem with the first possibility is that the Pentagon report officially rules it out. Plus, why would the US endanger and hinder their own navy by testing new technology in restricted training areas? And why invest so much money and resources investigating something they already know is theirs? The problem with the second option is that almost every advanced country in the world has also had, or still has, a UFO program like AATIP.[11] They are also trying to figure out what UFOs are. Plus, if a country had such a huge technological advantage, it would surely have already been displayed in the field of combat. Option three is an absolute last resort that should only be considered when all other possibilities have been exhausted.

One of the biggest questions mankind has asked is, *Are we alone?* Given the recent revelations from the Pentagon and the increased amount of newly available data, now is the time to scientifically investigate whether the UFO phenomenon can shed any light on this question. The answer to this question may have huge implications for all of mankind, or more prosaically, it may simply highlight a monumental failure in US national security. Either way, the stakes are high.

Given the possible implications, a thorough and careful investigation is crucial. Scientific integrity compels us to look at the data dispassionately and judge its validity without prejudice. We must let the data drive the narrative. As a scientist, one must be extremely discerning when it comes to the type and quality of data used in a study. A conclusion is only as strong as the evidence supporting it. For this reason, it is prudent to review the major types of evidence when it comes to UFO cases, and just how reliable each type of evidence is.

1.3 Filtering Evidence

Finding the truth is something like applying a series of filters. The first filter might be coarse, allowing most impurities to pass through. The second filter may be a little more refined, blocking all but the smallest granules. After passing through a series of successively finer filters, what remains is pure. Similarly, we must apply a set of filters when considering evidence. For example, most UFO incidents involve just one untrained eyewitness, with no other corroborating evidence. These cases may pass through the first filter but are blocked by the second. They are too unreliable. Some cases, however, involve several expert eyewitnesses with corroborating data from multiple independent sensors. These cases are more reliable, having passed through several truth filters.

1.3.1 Filter 1: Eyewitness Testimony

In 1985, Kirk Bloodsworth was convicted, and sentenced to death, for the rape and first-degree murder of Dawn Hamilton, a nine-year-old girl from Maryland. Kirk was convicted largely based on the testimony of five eyewitnesses. With the advent of DNA testing in the early 1990s, it was eventually possible to prove Kirk's innocence. In 1993, nine years after first going to jail, Kirk was released from prison. How could such a tragedy happen? How could five eyewitnesses all be so wrong? Unfortunately, such life-ruining mistakes are all too common. Since DNA testing became prevalent in the 1990s, 73 percent of the 239 convictions that were overturned because of DNA evidence were based on eyewitness testimony, according to one research body.[12]

But why is eyewitness evidence so unreliable? One key factor is the way the memory works. After recalling a memory many times, we no longer recall the actual event itself, we recall the last time we relived that memory.

The eminent psychologist Elizabeth F. Loftus of the University of California, Irvine, describes the process as "more akin to putting puzzle pieces together than retrieving a video recording."[13] If we were in a particularly bad mood when we last retrieved and recounted a memory, then we may put the puzzle pieces together in a slightly different way than if we were happy. Moreover, even the person interviewing the eyewitness can influence how the memory is reconstructed, for example by accidentally or deliberately implanting false memories.[14] There is a multitude of other uncontrolled variables that make untrained eyewitnesses particularly poor observers, such as heightened emotions and stress levels during the event, visual acuity issues, and unknown intoxication levels.

Generally, having multiple eyewitnesses to an event is considered stronger evidence than just a single eyewitness. This is especially true if the eyewitnesses independently observed the event and did not communicate with one another until after they had recorded their testimony. It is important to consider such fine details, as eyewitnesses are known to influence each other, either through peer pressure, accidentally implanting false memories, or other means.[15] Crucially, independent witnesses are less likely to be influenced by the same physical conditions. For example, numerous witnesses all located at the same site may experience the same atmospheric optical illusion. However, if an event is observed independently from many different angles and viewpoints, then optical illusions are far less likely to be reported by all observers. Nevertheless, as the example of Kirk Bloodsworth demonstrates, even multiple witnesses can be very wrong.

Some of these issues are mitigated by expert, trained observers. For example, fighter pilots typically have thousands of hours of flight experience under highly stressful conditions. They know the silhouette and flight capabilities of every aircraft in the sky. They know how to remain calm in stressful combat situations. Their visual acuity, general health, and intoxication levels are checked regularly and thoroughly. Fighter pilots make significantly better eyewitness observers, but they are not infallible.

A worrying example of this comes from a study by Professor J. Allen Hynek of Northwestern University. Hynek concluded that "Surprisingly, commercial and military pilots appear to make relatively poor witnesses." He determined that commercial pilots had an 89 percent misperception rate and military pilots had an 88 percent misperception rate.[16] Although Hynek did verify that a pilot's ability to identify objects familiar to them was excellent, such as aircraft and structures on the ground, they were particularly bad at identifying astronomical objects, such as stars and planets.

As before, having multiple independent trained witnesses is generally considered a stronger form of evidence than a single trained witness. For example, in the case of Commander David Fravor reporting the "tic-tac" UFO, the fact that this was also visually confirmed by three other pilots, including Lieutenant Commander Jim Slaight and Lieutenant Commander Alex Dietrich, two of which had an entirely different vantage point, strengthens the evidence significantly.

1.3.2 Filter 2: Single Sensor Data

Next in the evidence hierarchy is sensor data. Making quantitative measurements using precise experimental sensors essentially eliminates human error. Machines do not get stressed or report false memories. The data they record is preserved with high fidelity and can be made available for all to analyze. Moreover, the fact that the data is quantitative enables the use of powerful mathematical and statistical tools. This is hard to do based on the memory of an eyewitness.

Nevertheless, all experimental measurements come with some amount of associated error. Statistical errors can be reduced by performing many independent repeat trials, but this is hard to do for a single one-off event like a UFO incident. When considering data obtained from a *single sensor*, we must be acutely aware of potential systematic errors. For example, during the 2004 *Nimitz* encounter, the radar technicians knew that the most likely source of the weird radar signals was a fault with the radar system. Being diligent professionals, they rebooted, recalibrated, and ran extensive diagnostic tests on the entire radar system aboard the USS *Princeton*.

1.3.3 Filter 3: Multiple Sensor Data

When considering observations that appear consistently across *multiple sensors* our degree of confidence increases dramatically. For example, in the case of the 2004 *Nimitz* encounter, it is highly unlikely, although still possible, that the radar system aboard the USS *Princeton*, the radar system on the E-2 Hawkeye aircraft in the sky, and the infrared (FLIR) camera on the F/A-18F Super Hornet fighter jet would all simultaneously and consistently malfunction.

Even detecting a single event across multiple corroborating sensors still presents issues related to repeatability and reproducibility. One cannot perform multiple repeat trials on a single event in time. Nor can the scientific results from a single event be independently collected and verified by a separate group of researchers.

1.3.4 Filter 4: Physical Evidence

Physical evidence is the most reliable kind of evidence. However, we must be clear what we mean by physical evidence in this context. Here, I define physical evidence as tangible substance that can be analyzed a potentially indefinite number of times. For example, a large amount of blood found at a crime scene. This is a physical substance that can be repeatedly measured and even sub-divided for analysis by independent research groups. I distinguish this from the other commonly used meaning of physical evidence, namely evidence that is non-biological.

Relating this to the subject of this book, say we recovered a piece of an extraterrestrial craft, we could then perform all sorts of experiments and repeat the measurements an arbitrary number of times. Moreover, multiple scientific groups could perform independent measurements of the same subject by taking small samples.

It is important to realize that the nature of certain events may yield practically no physical evidence. For instance, imagine a small asteroid streaking across the night sky and completely burning up. In this case, since the asteroid completely disintegrated, there is very little tangible evidence remaining that the event occurred. Nevertheless, it did, and this could be proven beyond reasonable doubt via data from multiple sensors and expert witnesses.

1.3.5 Summary

In this book, we will not study any case that is based solely on eyewitness testimony, either from trained or untrained observers. Although I acknowledge that trained observers, such as Top Gun fighter pilots with thousands of hours of flight experience, are among the best type of observers when it comes to identifying aircraft, I must set the standard of evidence as high as possible. I want to be as sure as I possibly can about any conclusions reached. The specific cases selected for study are therefore those that are corroborated by single sensor data, multiple sensor data, or physical evidence.

There are some very compelling cases involving expert eyewitness testimony of sensor data. For example, in the 2004 *Nimitz* case, radar operator Senior Chief Kevin Day has gone on record as saying that radar data from the USS *Princeton* showed that the UFOs dropped from 28,000 feet down to sea level in just 0.78 seconds. Such a maneuver would require a staggering acceleration, far beyond what humans could withstand. However, this radar data remains classified and therefore unverifiable. So, in this case, we are essentially forced to rely on the memory of an expert eyewitness, even if it is a memory of actual sensor data. Unfortunately, I must demote such secondhand evidence to the filter one category. Therefore, I do not include such cases, despite their compelling nature.

To quantitatively evaluate the strength of each case, we will implement the following scoring system. The four categories of evidence we will evaluate are trained eyewitness testimony, single sensor data, multiple sensor data, and physical evidence. Each category will be evaluated based on the quantity and quality of the data, its self-consistency, and the reliability of its source. Each of these four subcategories will receive a whole number score out of 3, giving a total score out of 12 for each evidence type. The score from each evidence type will then be weighted by a factor based on the relative importance of that evidence type. Trained eyewitness testimony is considered the least important type of evidence and thus has a weighting factor of just one. The next most important is single sensor data, which receives a weighting factor of two. Multiple sensor data is weighted by a factor of three, and physical evidence by a factor of four. Each case will therefore receive a score out of a total of 120 (since $(1 \times 12) + (2 \times 12) + (3 \times 12) + (4 \times 12) = 120$) and will finally be expressed as a percentage. This scoring system is summarized in table 1.1.

TABLE 1.1
The scoring system for each case.

Evidence Type	Quantity	Quality	Consistency	Source	Total	Weighted Total
Eyewitness testimony	0–3	0–3	0–3	0–3	0–12	Total × 1
Single sensor data	0–3	0–3	0–3	0–3	0–12	Total × 2
Multiple sensor data	0–3	0–3	0–3	0–3	0–12	Total × 3
Physical evidence	0–3	0–3	0–3	0–3	0–12	Total × 4

1.4 Notes

1. Barack Obama, interview by James Corden. *Reggie Watts to Barack Obama: What's w/ Dem Aliens?* Los Angeles, CA, May 18, 2021).

2. Thomas S. Kuhn, 1970. *The Structure of scientific revolutions* (Chicago: University of Chicago Press, 1970).

3. Danish podcast, *Fingret, det pekande.* June 25, 2021 (accessed October 16, 2021, https://www.youtube.com/watch?v=MJB5B2ghemE).

4. US Department of Defence,. *Establishment of Unidentified Aerial Phenomena Task Force.* August 14, 2020 (accessed October 17, 2021, www.defense.gov).

5. The Nimitz Encounters, *Tic Tac Witnesses Kevin Day Interview*, October 12, 2019 (accessed October 17, 2021, https://www.youtube.com/watch?v=_2zRabdvKnw).

6. Helene Cooper, Ralph Blumenthal, and Leslie Kean, "'Wow, What Is That?'" *New York Times*, May 26, 2019.

7. Bryan Bender, "Senators Get Classified Briefing on UFO Sightings," *Politico*, June 19, 2019.

8. Office of the Director of National Intelligence, "Preliminary Assessment: Unidentified Aerial Phenomena," *dni.gov*, June 25, 2021 (accessed October 18, 2021, https://www.dni.gov/files/ODNI/documents/assessments/Prelimary-Assessment-UAP-20210625.pdf).

9. Bender, "Senators Get Classified."

10. Eli Rosenberg, *Former Navy Pilot Describes UFO Encounter Studied by Secret Pentagon Program*, December 18, 2017 (accessed October 18, 2021, https://www.washingtonpost.com/news/checkpoint/wp/2017/12/18/former-navy-pilot-describes-encounter-with-ufo-studied-by-secret-pentagon-program/).

11. Timothy Good, *Need to Know: Unidentified Flying Objects, the Military and Intelligence* (Cambridge: Pegasus Books, 2007).

12. Hal Arkowitz and Scott O. Lilienfeld, "Why Science Tells Us Not to Rely on Eyewitness Accounts," *Scientific American*, January 1, 2010.

13. Elizabeth F. Loftus, *Eyewitness Testimony: Civil and Criminal*, sixth edition (New York: Lexis Nexis., 2019).

14. Arkowitz and Lilienfeld, "Why Science Tells."

15. Loftus, *Eyewitness Testimony*.

16. J. Allen Hynek, *The Hynek UFO Report* (New York: Dell Publishing Company, 1977).

I
CASE STUDIES

2
Japan Airlines Flight 1628

UFOs are as real as the airplanes flying overhead.
—Paul Hellyer (Canadian Minister of National Defense)[1]

Acknowledgments: I would especially like to thank Kevin Knuth for his input on this case study. I also wish to thank Robert Powell and Peter Reali for their pioneering work on this case.

2.1 Case Description

THE FOLLOWING IS BASED on official Federal Aviation Administration (FAA) transcripts and documents available from the National Archives (*Identifier number 733667. Local identifier number 1203*).[2]

On November 17, 1986, Japan Airlines Cargo Flight 1628 (JAL1628) was flying over Alaska as part of a routine cargo flight. At the helm was Captain Kenju Terauchi, a former fighter pilot with nearly thirty years of experience and more than ten thousand hours of flight time. Also in the cockpit, were co-pilot Takanori Tamefuji and flight engineer Yoshio Tsukuba. The flight was relatively smooth and uneventful—that was until 5:11 p.m. local time.

At this point, the captain noticed two lights thirty degrees left of front and six hundred meters below. JAL1628 was cruising at an altitude of 10,600 meters with a ground speed of approximately 910 kilometers per hour. The strange pair of lights appeared to mimic the planes speed and direction exactly. The crew maintained visual contact with the UFOs for the next seven minutes. Then, they rapidly approached the cockpit, hovering just outside the window at only 150 to 300 meters.

The cabin lit up and the captain felt "the warmth of light" on his face. The two objects were now close enough to reveal distinct features. Both crafts were approximately fifty meters across, square in shape, and emitting amber jets of light. Each had a dark central rectangular body flanked on either side by a bright array of circular lights. A reprint of the original sketch made by captain Terauchi of the appearance of the craft can be seen in figure 2.1.[3]

Alarmed, the co-pilot scrambled a message to air traffic control (ATC) located in Anchorage, Alaska. The following is taken from the official transcript of communications between JAL1628 and ATC released by the Federal Aviation Administration (FAA). The speaker and time of each communication are printed on the left.

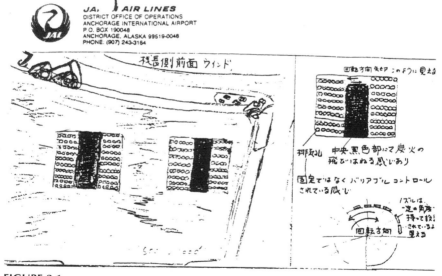

FIGURE 2.1
(Left) Captain Terauchi's sketch of how the two UFOs appeared from the cockpit window. The two craft flew in a one-on-top-of-the-other formation for several minutes before changing to the side-by-side configuration depicted. (Top, right) A sketch indicating that the middle section of the craft sparked a stream of lights that moved from left to right, then right to left. (Bottom, right) The captain's interpretation of what the craft may look like when viewed from above. *Credit*: National Archives

(5:19:15 p.m.) JAL1628: Anchorage Center, Japan Air sixteen twenty-eight; ah do you have any traffic, ah seven o'clock above?[a]
(5:19:32 p.m.) ATC: Japan Air sixteen twenty-eight heavy; negative.

Here, the term "heavy" refers to heavy radio interference. In a later interview, Captain Terauchi would state that whenever the "little ships" came close they would experience significant interference on their VHF radio when attempting communication with ground control. When pressed to describe the radio interference he heard, he said it was "some kind of, like, ah, jamming . . . it was just a noise, sounded like *zaa, zaa*." Through heavy interference, ground control manages to confirm that they do not see anything on the radar.

Nevertheless, the co-pilot continues to express his concern over what he is seeing:

(5:19:36 p.m.) JAL1628: Ah, Japan Air sixteen twenty-eight; roger and, ah we insight ah-two traffic-ah, in front of us one mile, about.[b]

Worried about a possible mid-air collision, ground control presses the crew to identify the UFO and suggest an immediate altitude change.

(5:21:19 p.m.) ATC: Japan Air sixteen twenty-eight heavy; Sir if your able to identify the type of aircraft, ah-and see if you can tell whether its military or civilian.
(5:21:35 p.m.) JAL1628: Ah, Japan Air sixteen twenty-eight; we cannot identify ah, the type, ah but, ah we can see, ah navigation lights and ah, strobe lights.
(5:21:48 p.m.) ATC: Roger sir, say the color of the strobe and beacon lights?
(5:21:56 p.m.) JAL1628: The color is ah—white and yellow, I think.

At this point, it is important to note that the Japanese crew of JAL1628 were not fluent in English. In interviews following the event, with a translator present, the colors of the strobe and beacon lights were described as yellow, amber, and green. Yellow and amber are not conventional colors for aircraft.

Shortly after this, radio communications underwent extremely heavy interference, forcing a change of radio frequency. Then, a few minutes later, the crew report the UFO vanished and radio communications return to normal. ATC also confirm that they no longer see the UFO on the radar.

ATC then contact the Regional Operations Command Center (ROCC), a military facility run by NORAD, to see if they can see it on their radar.

---

a. The co-pilots reference to the seven o'clock direction is inconsistent with the UFO being in front of the cockpit. This may be a verbal or transcript error since only seconds later the co-pilot asks about traffic in front.

b. The distance reported by the co-pilot of about one mile is significantly less than that given by the captain in a later statement of approximately 500 to1000 feet (150 to 300 meters).

(5:23:37 p.m.) ATC: Ya, could you [ROCC] look ah, approximately forty miles south of Fort Yukon, there should be a code up there of one-five-five-zero. Can you tell me you see a primary target about his position?

While waiting for the ROCC to get back to them, the Anchorage ground control contact JAL1628 once again.

(5:24:50 p.m.) ATC: Japan Air sixteen twenty-eight; do you still have ah visual contact with the ah traffic?
(5:24:53 p.m.) JAL1628: Affirmative, ah so we radar contact ah [unintelligible].
(5:24:50 p.m.) ATC: Japan Air sixteen twenty-eight heavy; roger sir I'm picking up a ah hit on the radar approximately five miles in trail of your six o'clock position, do you concur?
(5:25:12 p.m.) JAL1628: Ah negative, ah eleven o'clock ah eight miles ah same level over.

This indicates that a UFO was detected both by the radar onboard JAL1628 and by the ground radar at Anchorage at about the same distance from JAL1628. However, there is clear disagreement on its position relative to JAL1628, with ATC observing the six o'clock position and JAL1628 the eleven o'clock position. This may be explained by the captain's observation that the UFO would move rapidly around the plane in a circle. The captain's sketch of how the UFOs moved around the plane and the onboard radar display can be seen in figure 2.2.[4]

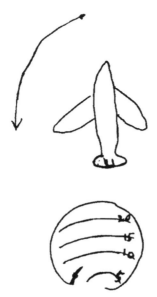

FIGURE 2.2
(top) Captain Terauchi's sketch of how the UFOs moved in a circle around the plane. (bottom) The captain's sketch of the onboard radar display, showing the UFO echo in the lower left of the screen at a range of about eight miles. In an interview on January 2, 1987, Captain Terauchi stated that the radar return from the UFO onboard JAL1628 was green in color, indicating a weak return. *Credit*: National Archives

The ROCC finally gets back in contact with ATC after attempting to locate the UFO on military radar:

> (5:25:43 p.m.) ROCC: Okay, I've got your squawk. It looks like I am getting some surge, primary return, ah I don't know if it's erroneous or whatever but . . .
> (5:25:50 p.m.) ATC: Negative, uh-uh, it's not erroneous. I want you [ROCC] to keep a good track on there, and if you pick up a code, and verify that you do not have any aircraft operating in that area military.
> (5:25:57 p.m.) ROCC: That is affirm. We do not have anybody up there right now, ah. Can you give me the position of the primary you're receiving?
> (5:26:03 p.m.) ATC: Okay, I'm not. I'm ah, picking up a primary—approximately five zero miles, south, right up there—right in front of the ah, one five five zero code.
> (5:26:18 p.m.) ROCC: Okay, I've got him about his-ah, oh-it looks like about, ah ten o'clock, at about that range, yes.
> (5:26:25 p.m.) ATC: Alright keep an eye on that, and ah-see if-ah, any other military [unintelligible] in that area.

This establishes that the ROCC run by NORAD also detected the UFO on radar. According to these official transcripts, the UFO has now been detected by three independent radar systems. In addition, the ROCC confirm that they do not have any military aircraft in the area and corroborate the approximate direction and range of the UFO reported by JAL1628. Referring to the UFO detected by the ROCC, ATC continues to remark:

> (5:27:49 p.m.) ATC: That is unknown to us.
> (5:27:50 p.m.) ROCC: It is what?
> (5:27:52 p.m.) ATC: It is, ah, unknown to us.
> (5:27:58 p.m.) ATC: Okay, we're not working that aircraft [unintelligible].

Let's just pause to understand what this means. Standard aircraft carry what is known as a transponder, which is a device that emits a specific radio response when it receives a pulse of electromagnetic energy sent out by a radar. Aircraft transponders usually emit a unique code that helps identify them to air traffic control. The ATC did not receive an identification code from the UFO, making it an unidentified flying object. The ATC received only unidentified radar returns from the UFO, meaning the electromagnetic energy emitted from the ground radar reflected from some kind of physical object, but one without a working transponder.

The Japan Airlines flight continued above the Eielson Air Force Base near Fairbanks, Alaska. The bright city lights of Fairbanks produced a pale glow. Captain Terauchi reports looking back at the pale light behind them and seeing the "silhouette of a gigantic spaceship." The captain would later estimate the huge UFO to be about two times the size of an aircraft carrier, making it about six hundred meters in length. Co-pilot Tamefuji reported that he could not see the UFO because his field of view was obscured because of the position of his seat in the cockpit. A sketch of what the captain observed can be seen in figure 2.3.[5]

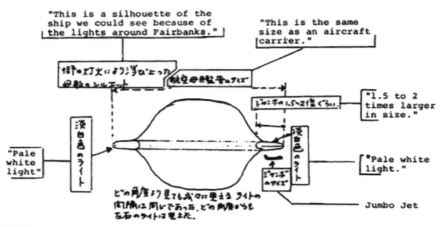

FIGURE 2.3
Captain Terauchi's sketch of the gigantic, silhouetted craft, including translated annotations. *Credit*: National Archives

A sense of panic gripped the crew, air traffic control was contacted immediately.

> (5:30:23 p.m.) JAL1628: Request ah deviate, ah ah from ah from object, request heading two four zero.
> (5:30:31 p.m.) ATC: Japan Air sixteen twenty-eight; roger fly heading two four zero. Japan Air sixteen twenty-eight ah heavy deviations approved as necessary for traffic.
> (5:30:49 p.m.) JAL1628: It's ah, quite big.
> (5:30:52 p.m.) ATC: Japan Air sixteen twenty-eight heavy; you're still broken say again.
> (5:30:56 p.m.) JAL1628: It's ah, I think ah, very quite big ah, plane.

For the next few minutes, the UFO followed the tail of JAL1628, despite substantial evasion efforts, including a 40-degree banking maneuver, an altitude drop from 35,000 to 31,000 feet, and a full 360-degree turn. The flight path of JAL1628 and its various maneuvers is reconstructed from the available radar data and can be seen in figure 2.4.[6] After being followed for approximately eight minutes, the crew of JAL1628 report that they no longer have visual contact with the UFO.

(5:39:04 p.m.) JAL1628: It ah—disappeared. Japan Air sixteen twenty-eight.

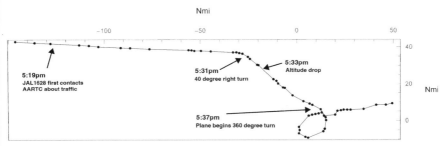

FIGURE 2.4
Key events during the flight of JAL1628.

But then, just a few seconds later, the ROCC military radar confirm a primary return for the UFO directly behind JAL1628.

(5:39:58 p.m.) ROCC: Ya, this is one dash two again. On some other equipment here we have confirmed there is a flight size of two around your one five-five-zero. Squawk one primary return only.
(5:40:05 p.m.) ATC: Okay, where is, is he following him?
5:39:58 ROCC: It looks like he is, yes.
(5:40:23 p.m.) ATC: Japan Air sixteen twenty-eight; roger. Sir, the military radar advises they do have a primary target in trail of you at this time.

The air traffic controllers at Anchorage are now becoming increasingly concerned, concerned enough to offer a rapid military response from Eielson Air Force Base and issue a second warning.

(5:40:13 p.m.) ATC: Roger sir. Would you [JAL1628] like our military to scramble on the traffic?
(5:40:17 p.m.) ATC: Negative, negative.
(5:40:35 p.m.) ATC: Japan Air sixteen twenty-eight heavy. Military radar advises they are picking up intermittent primary target behind you in-trail, in-trail I say again.

The captain immediately turned down the offer of military intervention. In his testimony after the event, Captain Terauchi would say, "I knew that in the past there was a US military fighter called the Mustang that had flown up high for a confirmation and a tragedy had happened to it. Even the F-15 with the newest technology had no guarantee of safety against the creature with an unknown degree of scientific technology." This appears to be a reference to the Mantell case of 1948, suggesting the captain had a prior interest in UFOs.[7]

A few minutes later, the UFO reappeared.

> (5:42:35 p.m.) JAL1628: Ah we have—Anchorage Center Japan Air one six two eight; ah we have in sight same position over.
> (5:42:35 p.m.) ATC: Japan Air sixteen twenty-eight understand in sight in same position.

Air traffic control contacted the nearby United Airlines Flight 69 (UA69) at 5:44:43 p.m., which was approximately 110 nautical miles away from JAL1629 at the time. The controller asked whether UA69 would divert and approach JAL1628 to visually confirm the trailing UFO, to which the UA69 captain agreed. As UA69 got within about twenty-five nautical miles of JAL1629, the crew of JAL1629 reported that the trailing UFO began to drop further behind.

> (5:48:34 p.m.) JAL1629: Ah-now, ah-ah-moving to ah-around ten mile, now-ah-ah-position-ah-seven, ah-eight o'clock, ten mile.

After about seven minutes UA69 had approached to within twelve nautical miles of JAL1628.

> (5:51:32 p.m.) UA69: Ah, Center from United ah sixty-nine. Ah-the-ah-Japan Airliner is silhouetted against a-ah-light sky. I don't see anybody around him at all. I can see his contrail, but I sure don't see any other airplanes. Do you see him?

By this time, JAL1628 had also reported to the ATC that the UFO had disappeared from visual and radar contact.

At 5:52:28 p.m., ATC contacted TOTEM 71, a military non-combat aircraft that was already in the sky, for a second attempt at an independent visual confirmation of the UFO. TOTEM 71 visually observed JAL1628 for the next few minutes, without seeing any UFO near it.

JAL1628 continued to Anchorage without further incident, landing safely at 6:20 p.m. A few months after the incident, Captain Terauchi was "grounded" and moved to a desk job by Japan Airlines, supposedly for talking to the press about the incident. Several years later, he would be reinstated as a pilot, before eventually retiring in Kanto, Japan.

Japan Airlines Flight 1628

2.2 Analysis

The following analysis is based on the computer printout of continuous radar data from Anchorage Air Traffic Control as provided by the FAA and available from the National Archives (Identifier number 733667. Local identifier number 1203).[8]

2.2.1 How to Track a UFO on Radar

In this subsection, we outline the methodology we will use in the analysis of the incident involving Japan Airlines Flight 1628.

The data analyzed was collected by Anchorage Air Traffic Control on November 17, 1986, between approximately 5:16 p.m. and 5:47 p.m. local time.[c] The radar in operation during the incident was the AN/FPS-117 long-range 3-D phased array antenna radar system. This radar has an uncertainty in range measurements of only fifty meters, and a scan rate of five revolutions per minute (RPM), which equates to a complete scan of the airspace every twelve seconds.[9] An illustrative sample of the radar data recorded by Anchorage Air Traffic Control during the incident is shown in figure 2.5.[10]

The available data contains four types of radar return: primary, secondary, correlated, and uncorrelated. Primary radar returns are pulses of electromagnetic energy emitted by the ground radar that are reflected from the surface of the aircraft. The time delay between sending and receiving the reflected pulse is a measure of the range. Secondary returns come from an aircraft's

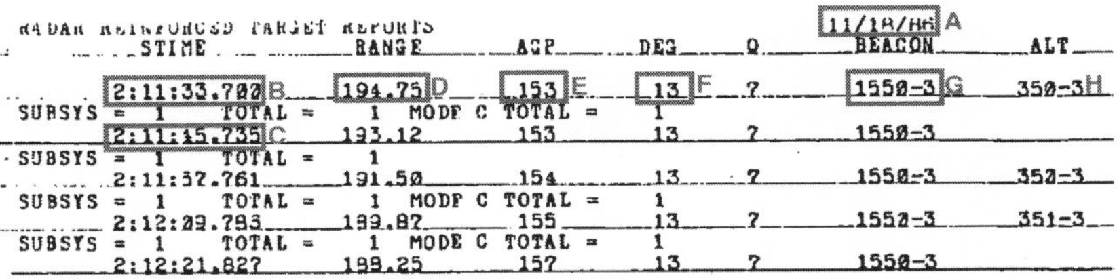

FIGURE 2.5
Sample radar data captured by Anchorage Air Traffic Control during the incident involving Japan Airlines Flight 1628. Key features of the data are labeled with letters; A) The date the radar data was printed; B) and C) The time of successive radar sweeps in Universal Coordinated Time (UTC); D) The distance of the object from the radar in nautical miles; E) and F) The direction of the aircraft in the horizontal plane; G) The unique aircraft identifier code, which was 1550 for JAL1628 and 1200 for the UFO. H) The altitude of the object. *Credit*: National Archives

---

c. The radar data was reconstructed and printed the next day on November 18, 1986, as shown by the date in figure 2.5.

transponder, a device that reacts to the receipt of a radar pulse by emitting a return signal. The secondary signal emitted by a transponder is usually stronger than the weak primary reflection, typically making them more accurate and reliable sources of data. Furthermore, a code that uniquely identifies each aircraft can be added to the secondary transponder signal. An uncorrelated radar return occurs when the pulse sent up toward the aircraft returns off the surface of the aircraft at a slightly different moment than the transponder signal. In this case, the two signals do not appear to come from the same place, and hence are said to be uncorrelated. A correlated return occurs when the primary and secondary returns come from the same place, or at least they appear within the same computer radar cell.

In this study, the flight path of JAL1628 and the UFO are reconstructed using secondary radar returns, due to their increased reliability. In addition, the trajectory of the UFO has been corroborated using correlated returns. The radar data was converted into three-dimensional Cartesian coordinates using some simple trigonometry implemented in the computer program Mathematica, details of which can be found in the appendix at the end of this chapter, along with the equations of motion used to model the UFO's acceleration and maximum speed. The radar data is only analyzed between approximately 5:16 p.m. and 5:47 p.m. local time, since this time frame captures the salient features of the incident.

2.2.2 Remarkable Results

In this subsection the main results derived from the radar data during the incident involving Japan Airlines Flight 1628 are given.

We begin by overlaying the flight path of the UFO and JAL1628, as shown in figure 2.6. The airplane's flight path during the incident is shown in red, and the UFO's path in blue. Although the two paths appear to intersect, they are in fact at very different altitudes, with the altitude of JAL1628 typically between five and six nautical miles (Nm), while the UFO is typically found between one and two Nm above ground.

Figure 2.7 focuses solely on the UFO's flight path, from a perspective looking downwards from above (plan view). Figure 2.7 has some interesting features, namely a series of eleven successive "jumps" from the bottom-left to the top-right of the plot. Each of these jumps covers approximately seventy nautical miles in a matter of seconds. Another interesting feature is the concentrated cluster of movement in the top-right of the graph, with the UFO displaying controlled non-random movements focused on one specific area. In figure 2.8, we zoom-in on this concentrated cluster of movement, again from a plan view perspective. This cluster of movement was concentrated

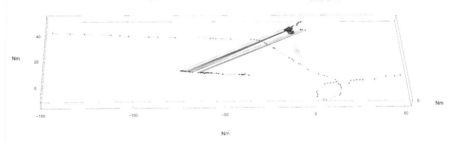

FIGURE 2.6
The combined flight paths of flight JAL1628 (lighter gray) and the UFO (darker gray). Each axis is given in units of nautical miles (Nm).

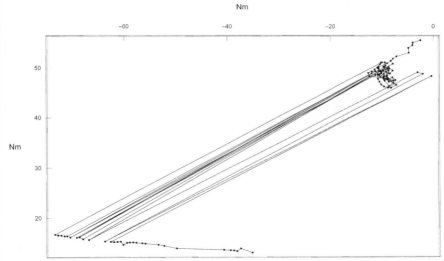

FIGURE 2.7
A plan view of just the UFO's flight path.

within a radius of only about five nautical miles. The flight path in this region does not appear to be entirely random, with a clear semi-circular arc and other geometric features.

In table 2.1, we present the maximum speed and the acceleration of the UFO for each of the eleven jumps. The maximum speed is given in multiples of the speed of sound in air, namely the so-called Mach number. When an object travels through air it sets up a series of pressure waves in front of it. As the speed of the object increases the waves compress together, unable to get out of each other's way, until eventually at the speed of sound the waves combine into a single shockwave—a sonic boom. An everyday example of a

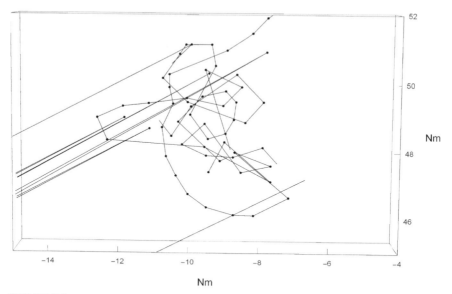

FIGURE 2.8
A plan view focusing on the concentrated cluster of movement of the UFO.

TABLE 2.1
A table of 11 jumps in the UFO's flight path and the maximum speed and acceleration of each. Maximum speed values are given in terms of multiples of the speed of sound (Mach number). Acceleration values are expressed in multiples of g, where $g = 9.81$ ms^{-2} is the acceleration on Earth due to gravity.

Jump	Max. Speed	Acceleration
1	53 ± 1	256 ± 3
2	83 ± 2	612 ± 10
3	52 ± 1	244 ± 3
4	53 ± 1	256 ± 3
5	322 ± 29	9,656 ± 815
6	77 ± 2	548 ± 10
7	52 ± 1	252 ± 3
8	76 ± 2	536 ± 10
9	316 ± 29	9,450 ± 815
10	76 ± 2	536 ± 10
11	350 ± 29	10,786 ± 815

sonic would be the crack of a whip as the tip breaks the speed of sound. As can be seen from table 2.1, the UFO had a maximum speed of between Mach 52 and Mach 350, far exceeding the speed of sound barrier on at least eleven occasions. Strangely, not a single sonic boom was reported during the incident. To put these speeds in context, the fastest known manned aircraft, the North American X-15, has a top speed of only Mach 6.7.

The acceleration of the UFO for each of the eleven jumps is given in the third column of table 2.1, expressed in multiples of g, where $g = 9.81 \text{ms}^{-2}$ is the acceleration on Earth due to gravity. These acceleration values are illustrated graphically in figure 2.9 and figure 2.10. The equations of motion used to determine the maximum speed and acceleration of the UFO during the eleven maneuvers are described in the appendix at the end of the chapter, along with details on how the errors are computed.

Let's put figure 2.9 and figure 2.10 in perspective. Fighter pilots in rapid vertical climbs can reach accelerations of around 8 to 9 g. Even at this level, special compression suits are required to stop blood draining from the head and rendering pilots unconscious. Modern military aircraft can maintain structural integrity up to a limit of around 15 g. For example, the F-16 Fighting Falcon can withstand 9 g when fully loaded with fuel, and the F-35 Lightning II can withstand 13.5 g.[11] Even missiles begin to fail at around 35 to 50 g.[12] Humans can endure surprisingly large accelerations, but only for very short durations. For instance, the world record is held by Colonel John Stapp, who endured an acceleration of 46.2 g for about a second.[13] Stapp's record came at a price, however. The blood vessels in his eyes burst open due to the extreme g-forces, making him temporarily blind. His ribs were cracked and both wrists were broken, and his circulatory and respiratory systems were heavily damaged. Needless to say, he was in a bad way after experiencing this g-force for less than a second. The ability of the human body to survive high accelerations significantly decreases with increased time exposure, dropping to only 25 g for durations over one second.[14] A few lucky racing car drivers have survived impacts that exceeded a peak of 100 g for just an instant, up to the highest ever recorded non-fatal peak force of 214 g, as measured by the accelerometer onboard an Indy Car during a high-speed crash.[15]

Figure 2.9 shows that the UFO, on eight separate occasions, underwent accelerations well above this highest survivable peak force, and for a significantly longer time. In figure 2.10, we see three UFO maneuvers that had accelerations of approximately 10,000 g, more than forty-four times greater than the maximum peak force a human has ever survived. It has been shown that the human skeleton can support a maximum vertical compressive force of about 90 g, beyond this our bones begin to quite literally be crushed into pieces.[16] At 10,000 g, a pilot would be reduced to a liquified mess on the wall. It is impossible for humans or any known aircraft to survive such extreme g-forces.

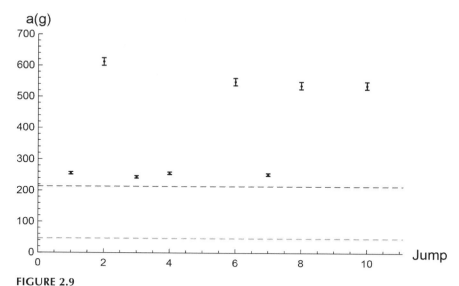

FIGURE 2.9
The eight UFO accelerations within the range (244–612) g. For reference, the dashed line at the very bottom indicates the point at which fighter pilots begin to lose consciousness (~10 g). The second dashed line from the bottom denotes Colonel John Stapp's world record for non-instantaneous g-force endured by a human (46.2 g). The upper dashed line indicates the highest recorded survivable instantaneous force (214 g).

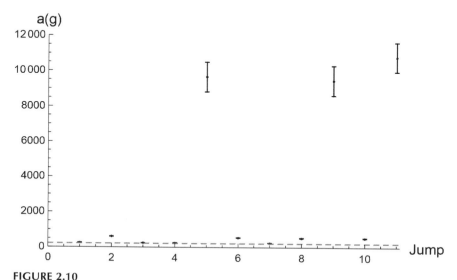

FIGURE 2.10
A rescaled version of figure 2.9, displaying the three highest recorded UFO accelerations within the range (9450-10786) g. The dashed line, indicating the highest peak force a human has ever survived, is barely visible on this scale.

2.2.3 Evaluation of Evidence

This subsection aims to evaluate and summarize the evidence in this case.

Eyewitness Testimony

Most events reported by Captain Terauchi were also corroborated by the co-pilot Takanori Tamefuji and flight engineer Yoshio Tsukuba. Since there were three trained observers in this case, I rate the quantity of trained eyewitnesses as 2 out of 3. A perfect score of 3 out of 3 would require a greater number of trained observers.

At the time of the incident, Captain Kenju Terauchi was a highly experienced pilot with nearly thirty years' experience and more than ten thousand hours of flight time. The co-pilot and flight engineer were also very experienced. However, there are some issues with the quality and consistency of the expert eyewitness testimony in this case. For example, when the co-pilot first contacted Anchorage ground control at 5:19 p.m. he inquired about traffic in the seven o'clock position, which is inconsistent with the claim that the UFO was directly in front of the cockpit at that time. This is likely a language or transcribing error as only seconds later the transcript shows the co-pilot asks about traffic in front. Another example is the color of the lights coming from the UFO. At 5:21 p.m., the crew describe the lights as white and yellow. Yet, in interviews after the event, with a translator present, the colors of the lights were described as yellow, amber, and green. The Japanese crew of JAL1628 were not fluent in English, which makes their testimony unclear and possibly unreliable in places. Moreover, Captain Terauchi's reference to a previous UFO case in the FAA transcripts suggests a prior interest in the phenomenon, which may have influenced his testimony. For these reasons, I only give a score of 2 out of 3 for the quality and consistency of the eyewitness testimony.

On January 22, 1987, the organization then known as the Scientific Investigation of Claims of the Paranormal (CSICOP), now known as the Committee for Skeptical Inquiry (CSI), published a press release titled "UFO Mystery Solved." In this article, CSICOP claimed that the trained eyewitnesses observed Jupiter and Mars. Let's examine the validity of this claim. On November 17, 1986, at 2:22 Universal Coordinated Time (UTC), Jupiter was above the horizon with an altitude of about 13 degrees and appeared in the Southeast direction, as viewed from Anchorage, Alaska.[17] Jupiter had a magnitude of about negative 2.6 at that time and thus would have appeared quite bright in the night sky. At the same time and place, Mars also appeared above the horizon with an altitude of about eleven degrees in the South-Southeast direction but was much dimmer than Jupiter. The Plane flew in the Southwest direction for most of the incident, and so Jupiter and Mars would have appeared at about ninety degrees and seventy degrees to the left of front, respectively, as viewed from the cockpit. The first sighting of the UFO made by the captain was at 30 degrees left of front, which disagrees with the position of Jupiter by about sixty degrees and Mars by about forty degrees.

Therefore, I find it unlikely that the highly-experience Captain could have been mistaken about the direction by such a large degree. However, it remains a slight possibility that he mistook Jupiter and Mars for a UFO during the initial sighting. However, Jupiter and Mars cannot suddenly change position in the sky and rapidly approach the front of the airplane's cockpit. Nor do they look anything like Captain Terauchi's sketch of the two aircraft. Nor can they reflect a radar pulse.

Single Sensor Detection

As indicated in the official FAA transcripts and confirmed in the computer printout of the radar data, Anchorage Air Traffic Control received numerous radar returns over approximately thirty minutes from an unidentified object. In total the radar data amounts to some 170 pages, and so I rate the quantity of single sensor data as 3 out of 3.

These radar returns indicate the presence of an unknown physical object in the vicinity of flight JAL1628 that performed maneuvers far beyond the limits of human endurance, and far beyond the structural capability of any known aircraft. For example, on three separate occasions, the UFO displayed an acceleration over 9,400 g (see figure 2.10). These results do not tell us what the UFO was, but they do tell us what it was not. Assuming the radar was not faulty, the tracked object could not have been any known man-made aircraft, either military or civilian, and it was certainly not piloted by a human. Neither could survive such extreme g-forces. Moreover, the single sensor data, if valid, indicates that the unidentified craft reached a maximum speed of more than fifty times the speed of sound on at least eleven separate occasions (on three of these occasions it exceeded three hundred times the speed of sound; see table 2.1). Yet not a single sonic boom was reported. Not by the crew of JAL1628, nor by any ground control operator. This feature of the incident is puzzling.

One possible explanation put forward by the FAA for these anomalous radar returns is that they were simply down to a radar malfunction, known as a split radar effect. Essentially, the FAA claimed that the pulse being reflected from JAL1628 itself and the one coming from JAL1628's transponder, which usually appear as very close blips of light on the radar screen, were erroneously displayed with a large separation, giving the appearance of two objects rather than one.

This explanation is possible but does not correlate with all the known facts. For example, the extra return did not appear with every radar sweep, and did not appear with the same separation, which one would expect if it were a systematic fault with the radar. Moreover, the official transcripts indicate that the UFO was also independently detected by the radar onboard flight JAL1628 and by the military radar operated by NORAD, one of the most advanced military command centers in the world. It's very unlikely, although still possible, to have this split radar effect plague three independent radar systems at the same time. In an official statement, one of the air traffic controllers also stated that they essentially never get a split radar effect in the region of the sky in which the incident occurred.[2] Nevertheless, given the possibility of such a malfunction, I downgrade the quality of single sensor data to a score of 2 out of 3. There are no obvious inconsistencies in this data, and so I give a 3 out of 3 for this category.

Multiple Sensor Detection

As indicated above, the official FAA transcripts of radio communications during the incident indicate that a UFO was also independently detected by the radar onboard flight JAL1628 (at approximately 5:24 p.m. local time) and by the military radar operated by NORAD (at approximately 5:25 p.m. local time) in a largely consistent manner. Despite this, the actual radar data from NORAD or JAL1628 has never been made available. The two aircraft (UA69 and TOTEM 71) that were re-routed to check out the claims of JAL1628 did not detect any UFO, either visually or using their onboard radar. Therefore, I rate the multiple sensor data as only 1 out of 3 for both quality and quantity.

Physical Evidence

There is no known physical evidence in this case.

Summary

The evidence of a truly anomalous arial phenomenon in the Japan Airlines Flight 1628 incident is summarized in table 2.2. The eyewitness testimony and single and multiple sensor data all originates from official FAA radar printouts and transcripts, and so the reliability of the data source is rated 3 out of 3 for these categories of evidence.

TABLE 2.2
A table summarizing the evaluation of evidence in this case.

Evidence type	Quantity	Quality	Consistency	Source	Total	Weighted total
Eyewitness testimony	2	2	2	3	9	9
Single sensor data	3	2	3	3	11	22
Multiple sensor data	1	1	2	3	7	21
Physical evidence	0	0	0	0	0	0

Therefore, we rate the evidence for the Japan Airlines Flight 1628 case as:

Final verdict: 52/120 (43 percent)

2.3 Appendix

2.3.1 Kinematics

Assuming constant linear acceleration we may use the well-known equation of motion

$$s = ut + \frac{1}{2}at^2, \tag{2.1}$$

where s is distance, u is initial speed, t is time, and a is acceleration.

In our specific case, s is the total distance the UFO covered in the time between successive radar returns. The Euclidean distance between the start and end points (x_1, y_1, z_1) and (x_2, y_2, z_2) in three-dimensional Cartesian coordinates was computed using the Pythagorean expression

$$s = \sqrt{(x_2 - x_1)^2 + (y_2 - y_1)^2 + (z_2 - z_1)^2}. \tag{2.2}$$

By assuming the UFO was at rest before and after each jump,[d] we can simplify our equation of motion by setting u = 0.

---

d. This assumption is supported by the radar tracks depicted in figure 2.7, since the UFO performs successive jumps in opposite directions, which requires the object to momentarily stop and reverse its direction. We have also ignored the acceleration of the object due to gravity.

A lower bound on the acceleration is obtained by assuming that the UFO accelerated at a constant rate a for the first half of the total distance $s/2$, and then decelerated at the same rate for the second half. This procedure ensures the most conservative estimate of the UFO's acceleration.[18] The motion of the UFO is therefore modelled via

$$a = \frac{2\left(\frac{s}{2}\right)}{\left(\frac{t}{2}\right)^2} = \frac{4s}{t^2}. \tag{2.3}$$

The maximum speed is reached at the midway point, since the object continuously accelerates over the first half, before decelerating over the second half. Using simple kinematics, it can be shown that the maximum speed v_{max} in this scenario is then given by

$$v_{max} = \frac{1}{2}at. \tag{2.4}$$

2.3.2 Processing the Radar Data

The conversion of the raw radar data into Cartesian coordinates is described in this subsection.

The parts of the radar data we take as input are the RANGE, ACP, and ALT values. The RANGE gives the distance in nautical miles between the ground-based radar antenna and the airborne object being tracked. As the antenna rotates it generates a certain number of pulses known as Azimuth Change Pulses (ACP). The antenna used in this incident generates 4,096 pulses per 360-degree revolution, which equates to an angle of approximately 0.088 degrees per pulse. The ACP counter is reset by a reference pulse as the antenna rotates through a specific point each revolution. The ACP number for each radar return can therefore be used to determine the angle in the horizontal plane (azimuth angle) between the object being tracked and the reference point. The ALT column specifies the altitude of the object in units of feet divided by one hundred, so for example an ALT value of 350 equates to an altitude of 35,000 feet. We convert the ALT values into a declination angle, which is the angle of the object above the horizon in the vertical plane. All three values are then converted from spherical coordinates to Cartesian coordinates via the Mathematica function *FromSphericalCoordinates*. An illustrative example of the Mathematica code used can be seen in figure 2.11.

```
(***************** Enter input from radar data ******************)
In[13]:= RANGE = 55.75 ;
In[14]:= ACP = 995 ;
In[15]:= ALT = 74 ;
In[16]:= (***************** Compute the azimuth angle in radians ******************)
In[17]:= azimuthRad = (180 - (ACP * 0.088 )) * (3.1416 / 180) ;
In[18]:= (***************** Compute the declination angle in radians ******************)
In[19]:= opp = ALT * 100 * 0.0001646 ;
In[20]:= decRad = (3.14159 / 2) - ArcSin [opp / RANGE ] ;
In[21]:= (***************** Convert spherical coordinates into Cartesian coordinates ******************)
In[22]:= urbl = FromSphericalCoordinates [{RANGE , decRad , azimuthRad }] ;
```

FIGURE 2.11
An illustrative example of the Mathematica code used to convert the radar data into three-dimensional Cartesian coordinates, which is used to reconstruct the flight path of JAL 1628 and the UFO.

2.3.3 Error Estimates

Here, we outline how the errors for each variable are calculated.

Given the equation

$$a = \frac{4s}{t^2}, \qquad (2.5)$$

the total error in the acceleration Δa can be estimated via the propagation of error equation[19]

$$\Delta a = \sqrt{\left(\frac{\partial a}{\partial s}\Delta s\right)^2 + \left(\frac{\partial a}{\partial t}\Delta t\right)^2} = \sqrt{\left(\frac{4}{t^2}\Delta s\right)^2 + \left(-\frac{8s}{t^3}\Delta t\right)^2}, \qquad (2.6)$$

where Δs is the error in measurements of distance s and Δt is the error in measurements of time t. Based on the published specifications of the AN/FPS-117 radar system, we implement the error $\Delta s = 50$m. The error budget for time measurements made by the radar is estimated by averaging over variations in the time it takes the antenna to return to the same ACP number, namely the time it takes the antenna to complete a full rotation, which according to the radar specifications should be exactly twelve seconds. Randomly sampling three sets of these time measurements yields a mean average error of $\Delta t = 0.046$ seconds, which we conservatively estimate as $\Delta t = 0.1$ seconds to ensure we do not underestimate the error.

Likewise, given the equation of motion

$$v_{max} = \frac{1}{2}at, \qquad (2.7)$$

we can determine the error in the maximum speed v_{max} via

$$\Delta v_{max} = \sqrt{\left(\frac{\partial v_{max}}{\partial a}\Delta a\right)^2 + \left(\frac{\partial v_{max}}{\partial t}\Delta t\right)^2} = \sqrt{\left(\frac{t}{2}\Delta a\right)^2 + \left(\frac{a}{2}\Delta t\right)^2}. \qquad (2.8)$$

Here, $\Delta t = 0.1s$, and Δa is computed for each jump using equation 2.6.

2.3.4 Data

As a supplement to table 2.1, you can find some additional details on the eleven jumps performed by the UFO in table 2.3.

TABLE 2.3
A table showing the time t in seconds, the distance s in meters, the maximum speed in ms^{-1} and the acceleration in ms^{-2}.

Jump	Time (±0.1s)	Distance (±50m)	Max. Speed (ms^{-1})	Acceleration (ms^{-2})
1	14.5	130,877.5	180,860 ± 202	2,500 ± 35
2	9.4	132,695.8	28,229 ± 603	6,006 ± 100
3	14.7	130,391.4	17,717 ± 202	2,407 ± 35
4	14.3	127,764.1	17,893 ± 202	2,506 ± 35
5	2.3	126,912.0	109,643 ± 10,081	94,724 ± 7,995
6	9.8	127,865.6	26,176 ± 603	5,358 ± 100
7	14.1	123,601.7	17,540 ± 202	2,489 ± 35
8	9.8	125,599.5	25,709 ± 603	5,262 ± 100
9	2.3	124,748.7	107,542 ± 10,081	92,708 ± 7,995
10	9.8	125,466.0	25,683 ± 603	5,258 ± 100
11	2.3	134,149.1	119,137 ± 10,081	105,806 ± 7,995

2.4 Notes

1. Paul Hellyer, *Light at the End of the Tunnel: A Survival Plan for the Human Species* (Bloomington, IN: AuthorHouse, 2010).
2. Federal Aviation Administration (FAA), *Japanese Airlines JAL 1628 UFO Encounter, November 17, 1986*, n.d. (accessed October 19, 2021, https://www

.theblackvault.com/documentarchive/ufo-case-japanese-airlines-jal1628-november-17-1986/).

3. FAA, *Japanese Airlines*.
4. Ibid.
5. Ibid.
6. Wolfram Research, Inc., *Mathematica Version 10.2* (Champaign, IL, 2015).
7. Thomas F. Mantell, "Flier Dies Chasing a 'Flying Saucer,'" *New York Times*, January 9, 1948
8. FAA, *Japanese Airlines*.
9. Christian Wolff,. *Radar Tutorial Card Index of Radar Sets—Air Defense Radar, an/fps-117 "seek igloo" (rrp-117)*, n.d. (accessed October 19, 2021, https://www.radartutorial.eu/19.kartei/02.surv/karte007.en.html).
10. FAA, *Japanese Airlines*.
11. USAF, *F-16 Fighting Falcon*, n.d.(accessed October 21, 2021, https://www.af.mil/About-Us/Fact-Sheets/Display/Article/104505/f-16-fighting-falcon/).
12. Kevin H, Knuth, Robert M. Powell, and Peter A. Reali, "Estimating Flight Characteristics of Anomalous Unidentified Aerial Vehicles," *Entropy* 21, no. 10 (2019).
13. John Paul Stapp, *New Mexico Museum of Space History*, n.d. (accessed October 21, 2021, https://www.nmspacemuseum.org/inductee/john-p-stapp/).
14. Dennis F. Shanahan (NATO), *Human Tolerance and Crash Survivability*, n.d. (accessed October 21, 2021, https://5b41d4f4-a-62cb3a1a-s-sites.googlegroups.com/site/2olykeioalimou/diaphora-gia-diktyo/EN-HFM-113-06.pdf?attachauth=ANoY7crTKy-VbOuRHYVnD7WQn6JbJxQM89jlCDXhj7m6DBugSy5j2rBYEH5ROzfAP21WHYgzW2GZRtTaJZAbIzbg4zmBRf2EjhE0gt2vrqHky7eV5E9iEcFg9e_S7fYpHMDQq).
15. Shanahan, *Human Tolerance*.
16. Nikola Poljak, Dora Klindzic, and Mateo Kruljac, "Effects of exoplanetary gravity on human locomotion ability." *The Physics Teacher* 57, no. 6 (2019): 378–81.
17. John Walker, *Your Sky—Sky Map*, n.d. (accessed October 19, 2021, https://www.fourmilab.ch/yoursky/custom.html).
18. Knuth, Powell, and Reali, "Estimating Flight Characteristics."
19. Gnyra, *Propagation of Uncertainty Calculator*, n.d. (accessed October 1, 2021, https://nicoco007.github.io/Propagation-of-Uncertainty-Calculator/).

3
The Brazilian Fragments

After looking into this, I came to the conclusion that there were reports—some were substantive, some not so substantive—that there were actual materials that the government and the private sector had in their possession.

—Harry Reid (senate majority leader)[1]

Acknowledgments: Special thanks goes to Peter A. Sturrock, Robert M. Powell, Michael D. Swords, Mark Rodeghier, and Phyllis Budinger for their efforts and foundational work referred to in this case.

3.1 Case Description

ON SEPTEMBER 13, 1957, a mysterious package landed on the desk of journalist Ibrahim Sued, a columnist for the Rio de Janeiro newspaper *O Globo*. Carefully, he eased it open and felt for the contents. First, he recognized the familiar caress of paper. Continuing to the bottom of the envelope, his fingers struck something unexpected. Three cold lumps of metal.

He took the rugged fragments out and placed them on his desk and stared at them. Confused, he snatched the paper from the envelope, hoping for an explanation. The letter read:

Dear Mr. Ibrahim Sued,

As a faithful reader of your column, and an admirer of yours, I wish to give you, as a newspaperman, a "scoop" concerning flying discs. If you believe that they are real, of course. I didn't believe anything said or published about them. But just a few days ago I was forced to change my mind. I was fishing together with various friends, at a place close to the town of Ubatuba, Sao Paulo, when I sighted a flying disc! It approached the beach at unbelievable speed and an accident, i.e., a crash into the sea, seemed imminent. At the last moment, however, when it seemed it was almost striking the waters, it made a sharp turn upward and climbed rapidly on a fantastic impulse. Astonished, we followed the spectacle with our eyes, when we saw the disc explode in flames. It disintegrated into thousands of fiery fragments, which fell sparkling with magnificent brightness. They looked like fireworks, despite the time of the accident, at noon, i.e., at midday. Most of the fragments, almost all, fell into the sea. But a number of small pieces fell close to the beach and we picked up a large amount of this material—which was light as paper. I am enclosing a sample of it. I don't know anyone that could be trusted to whom I could send it for analysis. I never read about a flying disk being found, or about fragments or parts of a disk that had been picked up. Unless the finding was made by military authorities and the whole thing kept as a top-secret subject. I am certain the matter will be of great interest to the brilliant columnist, and I am sending two copies of this letter—to the newspaper and to your home address.

From the admirer,
[the signature was illegible].[2]

At this point, it is important to note that the person that sent the letter has never been identified due to the illegible signature. Furthermore, despite extensive searches by the Brazilian representative of the Aerial Phenomena Research Organization (APRO), not a single corroborating eyewitness of the incident was ever found. Later, two independent researchers also tried to track down the witnesses of the incident by interviewing people at random in Ubatuba.[a] The only person that claimed to have relevant information was a local fisherman who remembered visitors from a nearby inland town recalling such an incident and showing pieces of metal.[3] The physicist and astronomer Professor Pierre Kaufmann of Mackenzie University in Sao Paulo analyzed and translated the original letter, which provides a picture that is at least vaguely consistent with the fisherman's recollection. Professor Kaufmann stated:

---

a. The two independent researchers were Dr Olavo Fontes, a local medical doctor, and Joao Martins.

The writer was definitely not a "local fisherman." The letter was written in very good Portuguese, and we may infer that the writer was well educated, and that he and some friends were visiting Ubatuba for a fishing vacation. A "local fisherman" would never have read Sued's social column in O Globo, and would probably not have read any newspaper. If he had, it would have been a Sao Paulo State newspaper.[4]

The next day (September 14, 1957) Sued broke the story in the society column of O Globo. The title of the newspaper article read, "A Fragment from a Flying Disk!" Dr. Olavo Fontes, a medical doctor and chief of the gastroenterology section of the National School of Medicine in Rio de Janeiro, read the article and immediately contacted the editor due to his interest in UFO reports. Dr. Fontes arranged to visit Sued in his apartment. The following description of what he saw is taken from a document that Dr. Fontes sent to APRO on November 30, 1957.

> I saw the samples sent by the unidentified correspondent—three small pieces of a dull gray solid substance that appeared to be a metal of some sort. Their surfaces were not smooth and polished, but quite irregular and apparently strongly oxidized. The surface of one of the samples was shot through with almost microscopic cracks. The surfaces of all samples were covered in scattered areas with a whitish material. These whitish smears of a powdered substance appeared as a thin layer. The fine, dry powder was adherent but could be displaced easily with the nail. Mr Sued said the material appeared to be lead at first sight—because of the gray color—but I could see that it could not be lead. The material was light ... almost as light as paper.[5]

Upon leaving Sued's apartment, Dr. Fontes took all three metallic fragments with him. Dr. Fontes would later label and refer to the individual fragments as samples 1, 2, and 3. Sample 1 was apparently never photographed.[6]

Dr Fontes gave a piece of sample 1 to Major Roberto Caminha of the Brazilian Army on November 4, 1957. Later, he also sent another piece of sample 1 to the Brazilian Navy, specifically to Commander J. G. Brandao. The military kept the samples and sent no reply. However, as reported by C. Lorenzen in her 1962 book on the subject

> During 1958 when Dr Fontes was in the midst of his investigation of the strange metal he was visited by two members of a Brazilian intelligence agency. These two individuals at first made veiled threats of what might happen to him if he continued his inquiry into matters that "did not concern him." When it became apparent that Fontes could not be coerced into silence, they appealed to his "better judgement" to cooperate with them and turn all his notes and the strange metal over to them. [Fontes declined this request.][7]

There were also similarly strange interactions with the US military. The Aerial Phenomenon Research Organization (APRO) tried to get a piece of sample 1 analyzed by the US Air Force. According to Lorenzen

> APRO submitted a portion of a sample to an Air Force spectrographic lab for analysis. An "emission spec" was requested. The following day the emission spectrograph operator reported that he had accidentally burned the entire sample without obtaining an exposed plate. He requested another sample. APRO declined.[8]

It is important to point out that much of this back story cannot be verified with any real degree of certainty. All that we know for sure is that these fragments appeared somewhere in Brazil in 1957. Nevertheless, the potential importance of the material itself should not be underestimated. Such physical evidence could, at least in principle, provide definitive proof of extraterrestrial technology. Because of their potential importance, the Ubatuba fragments have been independently analyzed by almost a dozen world-leading research facilities, with scores of collaborating scientists from prestigious institutes such as Stanford University and the Massachusetts Institute of Technology (MIT).[b] As we will see, these tests indicate the Ubatuba fragments are quite remarkable, especially given when and where they were found.

3.2 Analysis

3.2.1 Bulk Composition

The first attempt to determine the chemical composition of the fragments was conducted in Brazil from September to November 1957 at the Mineral Production Lab, a laboratory run by the Brazilian government. Sample 1 was analyzed independently by two scientists, a Dr. Barbosa conducted the first analysis, and a Mr. Texeira the second. Crucially, both scientists used the same piece of equipment, a Hilger mass spectrograph (model DMA 1-412).[9] The results from both tests found that the fragment was 100 percent pure magnesium, with absolutely no impurities whatsoever. Even today, it is not possible to create 100 percent pure magnesium, let alone back in 1957. If this result was correct, and the fragment really was perfectly pure magnesium, it may suggest an extraterrestrial origin. The laboratory report from Mr. Texeira states:

---

b. The Ubatuba fragments were also the only pieces of physical evidence investigated by the Colorado UFO Project (the Condon Committee), which was the only investigation of UFOs ever funded by the United States government.

The spectrographic analysis identified the unknown metal as magnesium (Mg), and showed it to be absolutely pure—as it can be concluded from the study of the spectrographic plate taken with the Hilger spectrograph. No other metal or impurity was detected in the sample analyzed; even the so-called "trace elements," usually found with any metal, were not present.[10]

This led Dr. Fontes to boldly declare that:

> On the basis of this evidence, it is highly probable the metallic chunks picked up on the beach near Ubatuba, in Sao Paulo, Brazil, are extraterrestrial in origin.[11]

However, there are at least two possible down-to-earth explanations for this apparently stunning result. First, it is possible that the Hilger spectrograph used in the analysis was simply not accurate enough to detect the impurities present. Second, the apparatus may have malfunctioned during both tests. Let's look at these possibilities in more detail. The Hilger spectrograph is known to be sensitive enough to easily detect impurities at the level of 1,000 ppm (0.1 percent) or greater.[c] Later tests, using more advanced methods, found that calcium and other impurities consistently appeared at a level greater than 0.1 percent in samples 2 and 3. Yet, these impurities were not detected in sample 1. So, either sample 1 had a significantly purer composition than samples 2 and 3, or the apparatus malfunctioned during both tests of sample 1. Assuming the Hilger spectrograph malfunctions once out of every hundred tests (which is likely an overestimate given this apparatus is a high-quality instrument),[12] and if the two measurements using this device were independent, then the probability of both tests 1 and 2 failing is one in ten thousand. Quite unlikely. However, since the same device was used in both tests it is entirely possible that there was some persistent problem with the apparatus, which could make the probability of repeated malfunction much higher.

By late 1957, sample 1 had almost entirely been used up by experimental procedures that irreversibly damaged the sample. The final piece of sample 1 was supposedly lost or mislabeled.[13] All subsequent analysis was therefore performed on samples 2 and 3, making it impossible to know for sure whether sample 1 was really 100 percent pure magnesium.

Driven by these interesting first results, the remaining fragments were analyzed again in September 1958, this time at Oak Ridge National Laboratories in Tennessee, United States. This second analysis used an advanced spectroscope to accurately determine the fragments elemental composition. The conclusion was that these samples were 99.8 percent pure magnesium.

---

c. Dr. Fontes even claims that the sensitivity of the spectrograph was as high as 1 ppm, although I cannot find evidence to support this claim.

In December 1961, the Dow Chemical Company analyzed the Ubatuba fragments using an electron beam probe. This time, the fragment was found to be 99.98 percent pure magnesium.

Between 1966 and 1968, the United States Air Force funded a committee made up of physicists, astronomers, psychologists, a chemist, and an electrical engineer to investigate the UFO phenomenon. Formally known as the University of Colorado UFO Project, it was often informally referred to as the Condon Committee, after its high-profile director, Edward Condon. In February 1968, the Condon Committee analyzed the Ubatuba fragments at the government's Alcohol and Tobacco Laboratory. Using neutron activation and gamma-ray spectroscopy the fragment was found to be 99.9 percent pure magnesium.[14]

The results of these five independent experiments are summarized in table 3.1.

TABLE 3.1
Summary of the bulk chemical composition of the Ubatuba fragments reported by four independent research groups. The average magnesium purity is calculated to be 99.94 percent.

Year	Instrument	Mg Purity (%)
1957(a)	Hilger Spectrograph	100
1957(b)	Hilger Spectrograph	100
1958	Grating spectroscope	99.8
1961	Electron probe	99.98
1968	Gamma-ray spectroscopy	99.9

From this data, the average magnesium purity of the Ubatuba fragments is calculated to be 99.94 percent. So, what does this result mean? Does it suggest the fragments have an extraterrestrial origin? Or are they just ordinary lumps of metal?

At the time of the alleged Ubatuba incident in 1957, there were only a few places on Earth that manufactured magnesium at an ultra-high purity of 99.9 percent or greater. One was the American Magnesium Corporation, which had been producing 99.9 percent pure magnesium since the 1920s via the electrolysis of magnesium oxide in fluoride melts.[15] Another, the Dow Chemical Company based in the United States, had been manufacturing 99.9 percent pure magnesium via the electrolysis of magnesium Chloride (obtained from sea water) since the early 1940s.[16] Commercial companies in England and Australia had also been producing such high-purity magnesium since the

1940s via the reduction of magnesium oxide. In conclusion, it is strange that such a pure fragment of magnesium (99.94 percent purity) was found on a remote beach in Brazil in 1957 (allegedly), however, a terrestrial origin cannot logically be excluded based on the purity of magnesium alone.

3.2.2 Impurities

The Dow Chemical Company was the first to identify specific impurities in the Ubatuba fragments in December 1961. Trace elements, including strontium at an abundance of thirty parts per million (ppm), were detected using an electron beam probe. Later, in February 1968, the Condon Committee also detected strontium impurities, but this time at a higher level of 500 ppm.[17] In 1986, Professor J. C. Lorin of the University of Paris used a Cameca SIMS instrument to analyze the fragments. Professor Lorin identified the presence of strontium at an abundance of 700 ppm. In the 1990s, the invention of a new device (the inductively coupled plasma mass spectrometer) dramatically improved our ability to identify minuscule trace elements. Two chips from the Ubatuba fragments were analyzed using this new device in the Elemental Research facility in Vancouver in 1997. The first fragment had a strontium abundance of 916 ppm, and the second 568 ppm. The evidence for strontium impurities in samples 2 and 3 of the Ubatuba fragments is strong and persistent over time, with the average abundance at around 550 ppm (~ 0.055 percent). Since the detected impurity level seems to have risen with time, probably due to more sensitive equipment being used, the true Sr impurity level may be greater than 550 ppm. These results are summarized in table 3.2.

TABLE 3.2
Measured strontium impurities in the Ubatuba fragments. Average Sr abundance = 543 ppm.

Year	Group	Sr abundance (ppm)
1961	Dow Chemical Company	30
1968	Condon Committee	500
1986	University of Paris	700
1997	Elemental Research facility	916
1997	Elemental Research facility	568

The presence of strontium in the fragments is hard to explain. Dr Craig, the chemist employed by the United States Air Force to investigate the Ubatuba fragment as part of the Condon Committee, stated that

The high content of Sr [strontium] was particularly interesting, since Sr is not an expected impurity made by usual production methods, and Dr Busk [Research Director of the Dow Metal Products Department of the Dow Chemical Company in Michigan] knew of no one who intentionally added strontium to commercial magnesium. In all probability, the strontium was added intentionally during manufacture of the material from which the sample came.[18]

Furthermore, Dr. Donald Beaman and Dr. Laurence Solaski of the Dow Chemical Company also stated that the strontium impurities must have been deliberately added, because they are not used in magnesium production.[19] Dr. Couling, of the Battelle Columbus Laboratories, has also expressed his surprise at the presence of strontium in the magnesium fragment, saying that "he did not know of any place in magnesium technology where strontium is used." In short, normal production methods of pure magnesium cannot easily explain the presence of strontium.[d]

So, as suggested by these experts it appears likely that strontium was deliberately added to the bulk magnesium. But why? Well, magnesium has a very low density making it ideal for aeronautical applications where weight considerations are paramount. However, magnesium's low tensile strength and susceptibility to corrosion and combustion has prevented its wider use in modern aeronautical engineering. Interestingly, there is some recent evidence to suggest that adding strontium to magnesium alloys increases hardness and yield strength.[20] Additionally, studies show that adding strontium to magnesium suppresses oxidation, yielding a material with a higher combustion temperature.[21] Perhaps, then, strontium was intentionally added to magnesium for aeronautical engineering purposes. Based on the scientific literature, it seems unlikely that such a procedure could have been implemented in 1957.[22]

Besides strontium, other impurities were also identified in the outer surface of the fragments, including barium, chromium, cobalt, niobium, palladium, selenium, and yttrium. Interestingly, these specific elements are usually found in a technological setting. For example, barium is used in sparkplugs and vacuum tubes.[23] Chromium, cobalt, and niobium are frequently used in jet engines and rockets.[24] Yttrium is used in radar systems and within alloys used in high-tech devices.

To summarize, the presence of strontium impurities inside such a pure sample of magnesium is the most puzzling fact. The appearance of other trace elements in the outer surface that are usually associated with technology is odd, but given the specific combination of elements, an exploding jet or rocket is a possibility.

---

d. Although Dow Chemical Company records indicate that experimental magnesium alloys were very occasionally made that contained some strontium impurities prior to 1957.

3.2.3 Isotopes

Isotopes are like red apples. On the outside, they all have essentially the same size, shape, and color. But if you cut them open the number of seeds in the core can vary significantly. Similarly, isotopes are atoms of the same element, meaning they generally have the same appearance and chemistry. But at the core of the atom, in the nucleus, different isotopes have a different number of neutrons. For example, magnesium has three naturally occurring isotopes, ^{24}Mg, ^{25}Mg, and ^{26}Mg, which contain twelve, thirteen, and fourteen neutrons, respectively. However, different isotopes of the same element always have the same number of protons, in this case, twelve.

Isotopes are very useful when trying to determine when are where an object originated from. For example, the ratio of the abundance of two magnesium isotopes in a sample can be measured. If the sample is terrestrial in origin this ratio should fall within a specific well-known range of values. Objects with a non-terrestrial origin, such as meteorites or moon rocks, are known to have an isotopic magnesium ratio that differs from the terrestrial range of values. This difference stems from the early development of the solar system and certain atomic decay processes.[25] An object that originated outside our solar system would vary even more from the terrestrial value.[26] Thus, isotopic ratios can give a strong indication of an object's origin.

On Earth, normal values for the ratios ^{25}Mg/^{24}Mg and ^{26}Mg/^{24}Mg are

$$\left(\frac{^{25}Mg}{^{24}Mg}\right)_{Terrestrial} = 0.1266, \left(\frac{^{26}Mg}{^{24}Mg}\right)_{Terrestrial} = 0.1394. \quad (3.1)$$

The uncertainty of each value can be estimated by sampling the isotopic ratios at various locations on Earth and from various sources.[27] Details of this uncertainty is estimated can be found in the appendix at the end of this chapter. We now need to know ratio of magnesium isotopes in the Ubatuba samples to test whether there is any tension with a terrestrial origin.

We will focus on three independent studies of the ratio of magnesium isotopes in the Ubatuba samples. The first study was conducted by Dr. Peter Sturrock of Stanford University in the spring of 1997 using a secondary ion mass spectrometer. The second isotopic analysis was performed two decades later, in 2017, by Cerium Laboratories in Austin, Texas, using a high resolution inductively coupled plasma mass spectrometer (ICPMS). The ICPMS Services in Cleveland, Ohio, was used for the third test in 2018, and were conducted by Dr. Arthur Varnes using a Thermo Scientific iCAP-Q inductively coupled plasma mass spectrometer. The results of these three studies, together with the normal terrestrial values, are given in table 3.3.

TABLE 3.3
A table comparing the isotopic ratios of magnesium typically found on Earth with those found in three separate laboratory tests of the Ubatuba fragments. The uncertainty of each value is given in the appendix at the end of this chapter.

Isotopic ratio	Terrestrial value	Sturrock 1997	Austin 2017	Cleveland 2018
$^{25}Mg/^{24}Mg$	0.1266	0.1259	0.1273	0.1254
$^{26}Mg/^{24}Mg$	0.1394	0.1373	0.1334	0.1369

As can be seen from these values, there is some disagreement between the isotope ratio found on Earth and the isotope ratios found in the Ubatuba fragments. But how big is this disagreement, exactly? Quantitatively, this tension can be measured by the number of standard deviations, σ For reference, in the social sciences, a result is considered significant at a level of 2σ or greater. In particle physics, however, discovery is not declared until at least the 5σ level has been reached. Based on the data in table 3.3, and the uncertainty of each value given in the appendix, the tension between the isotopic ratios in the Ubatuba fragments and those on Earth can be calculated. The tension for each study is summarized in table 3.4.[e]

TABLE 3.4
A table displaying the tension between normal terrestrial isotopic ratios and those found in three independent studies of the Ubatuba fragments. The tension is expressed in terms of the number of standard deviations σ.

Study	($^{25}Mg/^{24}Mg$) σ tension	($^{26}Mg/^{24}Mg$) σ tension
Sturrock 1997	0.17	0.72
Austin 2017	0.17	2.07
Cleveland 2018	0.29	0.86

Therefore, even the social sciences would not count most of these differences as statistically significant, and particle physics certainly would not. Yet, there is one significant deviation from normal terrestrial values, namely the ratio ($^{26}Mg/^{24}Mg$) found by the Austin 2017 study, with a significance of 2.07σ. This corresponds to about a 2 percent probability, or 1-in-52 chance, that the difference between these values is simply a statistical fluke. This result is interesting and warrants further study since it is just above the 2σ threshold. However, since both the Austin 2017 and Cleveland 2018 studies analyzed the same sample, it is possible that the Austin 2017 result is due to an experimental error or random statistical fluctuation.

---

e. See appendix 3.3.1 for details on how these values were determined.

The Brazilian Fragments

Comparing the second and third columns of table 3.4 suggests that it is the level of the isotope ^{26}Mg in the Ubatuba fragments that is mainly responsible for deviations from normal terrestrial values. Indeed, the Sturrock 1997 study found the ^{26}Mg abundance of 10.58 percent was significantly different from the precisely known standard terrestrial abundance of 11.01 percent. In fact, all three studies indicate a lower-than-expected ^{26}Mg abundance.[28]

3.2.4 Evaluation of Evidence

Here we evaluate and summarize the evidence in the case of the Brazilian Ubatuba fragments.

Eyewitness Testimony

Firstly, we must be very clear about the fact that the (single) eyewitness has never been identified. Also, the witness claimed to be a fisherman, and so we assume they have no specific training that may bolster their eyewitness testimony. Nevertheless, there are some consistencies between the eyewitness testimony and the physical evidence extracted from the Ubatuba fragment.

Tests conducted at the Oak Ridge National Laboratories in September 1958 by Dr. Ellison Taylor, Dr. Cyrus Feldman, Dr. T. A. Welton and Dr. Robert Gray, revealed small fissures within the magnesium crystalline structure. This suggests the fragment underwent oxidation at high temperatures. Evidence of high-temperature oxidation was also reported by an independent group of metallurgists based on inspections of the microscopic lattice structure of the fragments. Moreover, in 1978, Professor Robert E. Ogilvie of the Metallurgy Department at MIT analyzed the sample using an oxygen x-ray map. Professor Ogilvie confirmed the fragment had undergone high-temperature oxidation at some point in the past, writing that:

> The structure is indeed unusual. In my opinion it could only have been formed by heating the magnesium very close to its melting point in air. It would be necessary to hold the temperature for only a minute or so. This would produce an oxide coating on the material, which is clearly visible. Also, oxygen would diffuse down the grain boundaries, thereby producing the oxide network.[29]

Therefore, three independent research groups all confirm that the fragments underwent high-temperature oxidation, supporting the eyewitness account that they originated from an aerial explosion.

You may recall from high school that magnesium burns in air with a brilliant white light. This light is so intense that eye protection should be worn to prevent damage from powerful ultraviolet rays. Since we now know that the Brazilian fragments are almost pure magnesium, if they really did come from an exploding aerial vehicle then such an event would have been extremely bright. This fits with the eyewitness' description of the explosion as being of "magnificent brightness . . . like fireworks," even at midday.

Extensive tests also showed large amounts of sodium, calcium, manganese, and titanium, all of which are present in seawater.[30] However, a comparison with sand and seawater samples taken from beaches near Ubatuba yield ambiguous results. The abundance of sodium and calcium in the Ubatuba fragments is consistent with exposure to seawater and sand in this region; however, other surface impurities are less consistent, especially the level of iron.[31] The measured surface impurities are therefore partially compatible with the magnesium fragment being found on a beach near Ubatuba, as claimed by the eyewitness.

Since the eyewitness statement is largely consistent with the physical evidence, we give a rating of 2 out of 3 for the consistency of eyewitness testimony. However, the quantity, quality and source of the eyewitness statement is very weak, and so we must rate each of these three categories as 0 out of 3.

Single and Multiple Sensor Detection

As far as we know, no sensor systems detected the alleged flight path of the UFO or its subsequent explosion. Thus, the total score for this evidence type is zero.

Physical Evidence

Is there any physical evidence to suggest the Ubatuba fragments have an extraterrestrial origin? To answer this, we must first analyze the likely terrestrial origins of the fragments.

Magnesium is a lightweight metal, having only two-thirds the density of aluminum. Therefore, magnesium plays a critical role in improving the power-to-weight ratio of aircraft and spacecraft, as well as missiles, which increases range and fuel efficiency. Up until the 1950s, magnesium was used extensively in airplane construction. However, magnesium is known to be highly combustible, especially when it has a large surface area to volume ratio, for example when cut into thin strips or when in powder form. Could the Ubatuba fragments have come from an aircraft, missile, or spacecraft that combusted and then exploded? Not according to Oak Ridge National Laboratory, who categorically ruled out any type of exploding aircraft, missile, or pyrotechnics as the source of the fragment due to the specific form and properties of the magnesium.[32]

A meteorite is any piece of solid debris that comes from outer space and manages to reach the surface of a planet or moon. For example, a comet or asteroid fragment that does not completely burn up and disintegrate when entering Earth's atmosphere. The total mass of such objects striking Earth each year is currently estimated to be between ten million and one billion kilograms. Meteorites are composed of only eight or so elements, one of which is magnesium.[33] Could our mystery metal fragments have originated from a meteorite that burned up and exploded upon entering Earth's atmosphere? This seems unlikely based on a comparison between the ratio of specific isotopes of magnesium found in the Ubatuba fragment and in a reference meteorite, as shown in the work of Powell et al.[34]

It is also possible that the whole thing was a hoax. However, to fake this incident, the hoaxer would have to somehow obtain high-purity magnesium from one of only a handful of manufacturers on Earth at that time, add strontium impurities to the metal via some unknown manufacturing process, then expose it to high temperatures in the presence of oxygen and place it in or near seawater. Possible, but not very plausible.

Overall, tests performed on the physical evidence are high in both quality and quantity, regardless of whether the results indicate an extraterrestrial origin or not. Several independent research groups have tested the Ubatuba samples using a variety of different techniques and highly sophisticated equipment. Many of the scientists involved in these studies come from world-leading research institutes, such as Stanford University and MIT.

Summary

Our analysis focused on three aspects of the Ubatuba fragments: their bulk composition, their chemical impurities, and their isotopic ratios. The bulk composition of sample 1 was measured to be 100 percent pure magnesium, but this result may be due to equipment sensitivity or malfunction. On average, the bulk composition of the tested fragments across samples 1 through 3 is 99.94 percent pure magnesium. Although this purity is unusual given the time and place of its alleged origin, it could have been produced by commercial processes available at the time. Regarding impurities, the presence of strontium in the samples is especially hard to explain. The presence of technological metals is interesting but could easily have a prosaic explanation. The isotopic ratio ($^{26}Mg/^{24}Mg$) found by the Austin 2017 study exhibits a greater than 2σ tension with the normal terrestrial value. At least in the social sciences, this result constitutes a real statistical anomaly.

In short, it is hard to find a plausible terrestrial origin for the Ubatuba fragments given when and where they were found. Nevertheless, we cannot logically infer from this that they are extraterrestrial in nature. This case is frustrating, since there are some tantalizing pieces of physical evidence that deserve a closer look, but also a distinct lack of credible eyewitness testimony and sensor data makes it less compelling. Our evaluation of this case is summarized in table 3.5 and the final verdict is given directly below it.[f]

TABLE 3.5
A table summarizing the evaluation of evidence in this case.

Evidence type	Quantity	Quality	Consistency	Source	Total	Weighted total
Eyewitnesses	0	0	2	0	2	2
Single sensor data	0	0	0	0	0	0
Multiple sensor data	0	0	0	0	0	0
Physical evidence	3	3	3	1	10	40

Final verdict: 42/120 (35 percent)

3.3 Appendix

3.3.1 Uncertainty of Terrestrial Isotopic Ratios

We use the measured isotopic ratios $^{25}Mg/^{24}Mg$ and $^{26}Mg/^{24}Mg$ reported in Powell et al. for eight different terrestrial sources.[35] These include samples from Australian Olivine minerals, the Amazon River, the Johnson Space Center, the Dow Chemical Company, Italy, dolomite minerals, the North Atlantic, and the Dead Sea. The isotopic ratios for these eight samples are summarized in table 3.6.

TABLE 3.6
A table of isotopic magnesium ratios obtained from eight different terrestrial samples.

Ratio	Olivine	Amazon	Johnson	DOW	Italy	Dolomite	N. Atlantic	Dead Sea
$^{25}Mg/^{24}Mg$	0.1260	0.1261	0.1262	0.1268	0.1272	0.1275	0.1279	0.1286
$^{26}Mg/^{24}Mg$	0.1379	0.1383	0.1381	0.1392	0.1405	0.1408	0.1420	0.1434

---

f. As a suggestion for further study, it may be possible to date the Brazilian fragments using the method known as rubidium-strontium dating. For this we must know the abundance of the strontium isotopes ^{87}Sr and ^{86}Sr as well as the abundance of the rubidium isotope ^{87}Rb. The first two have already been measured. If we also knew the abundance of ^{87}Rb then we may be able to determine how old the Ubatuba fragments are.

The Brazilian Fragments

The uncertainty of the normal terrestrial isotopic ratios ^{25}Mg/^{24}Mg and ^{26}Mg/^{24}Mg are estimated using the corrected sample standard deviation

$$\sigma = \sqrt{\frac{\Sigma(x_i - \mu)^2}{N-1}}, \qquad (3.2)$$

where x_i are the individual ratio measurements, μ is the sample mean and N is the sample size. Using equation 3.2 and the data from table 3.6 gives the expected terrestrial values, and their respective uncertainties, as

$$\left(\frac{^{25}Mg}{^{24}Mg}\right)_{Terrestrial} = 0.1266 \pm 0.0009, \left(\frac{^{26}Mg}{^{24}Mg}\right)_{Terrestrial} = 0.1394 \pm 0.0020. \quad (3.3)$$

3.3.2 Uncertainty of Cleveland 2018 Isotopic Ratios

The work of Powell et al. implies the uncertainty for the abundance of ^{24}Mg measured by the Cleveland 2018 group is $\Delta(^{24}\text{Mg}) = 0.12$.[36] Likewise, $\Delta(^{25}\text{Mg}) = 0.30$ and $\Delta(^{26}\text{Mg}) = 0.15$. If z denotes the ratio ^{25}Mg/^{24}Mg, then the corresponding uncertainty Δz is given by the propagation of error equation

$$\Delta z = \left(\frac{\Delta x}{x} + \frac{\Delta y}{y}\right) z, \qquad (3.4)$$

where x is the abundance of ^{25}Mg and y is the abundance of ^{24}Mg, with Δx and Δy their respective uncertainties. A similar calculation yields the uncertainty for the ratio ^{26}Mg/^{24}Mg. Since the lab results of the Sturrock 1997 and Austin 2017 tests did not include standard error estimates, we assume that these uncertainties are equal in magnitude to that of the Cleveland 2018 study.

3.3.3 Sigma Tension

The sigma tension between two values $A \pm \delta a$ and $B \pm \delta b$ is determined in the following way. First, take the modulus of the difference between the central values, namely $|A - B|$. Next, assuming the uncertainties are independent and follow a normal distribution, the uncertainty in $|A - B|$ is given by

$$\sqrt{(\delta a)^2 + (\delta b)^2}. \qquad (3.5)$$

The sigma tension between $A \pm \delta a$ and $B \pm \delta b$ is then given by

$$\frac{|A - B|}{\sqrt{(\delta a)^2 + (\delta b)^2}}. \tag{3.6}$$

The results are given in table 3.7.

TABLE 3.7
A table of isotopic ratios of magnesium typically found on Earth and those found in three separate laboratory tests of the Ubatuba fragments, including uncertainty estimates for each value.

Isotopic ratio	Terrestrial value	Sturrock 1997	Austin 2017	Cleveland 2018
$^{25}Mg/^{24}Mg$	0.1266 ±0.0009	0.1259 ± 0.0040	0.1273 ± 0.0040	0.1254 ± 0.0040
$^{26}Mg/^{24}Mg$	0.1394 ±0.0020	0.1373 ± 0.0021	0.1334 ± 0.0021	0.1369 ± 0.0021

3.4 Notes

1. Ralph Blumenthal and Leslie Kean, *No Longer in Shadows, Pentagon's UFO Unit Will Make Some Findings Public*, July 23, 2020 (accessed May 14, 2022, https://www.nytimes.com/2020/07/23/us/politics/pentagon-ufo-harry-reid-navy.html).
2. Peter A. Sturrock, "Composition Analysis of the Brazil Magnesium," *Journal of Scientific Exploration* 15, no. 1 (2001).
3. Roy Craig, *UFOs: An Insider's View of the Official Quest for Evidence* (Denton: University of North Texas Press, 1995).
4. Sturrock, "Composition Analysis."
5. Ibid.
6. C. Lorenzen, *The Great Flying Saucer Hoax: The UFO Facts and Their Interpretation* (New York: William-Frederick Press, 1962).
7. Lorenzen, *Saucer Hoax*.
8. Ibid.
9. Sturrock, "Composition Analysis."
10. Lorenzen, *Saucer Hoax*.
11. Ibid.
12. Sturrock, "Composition Analysis."
13. Ibid.
14. E. U. Condon and D. S. Gillmor, *Scientific Study of Unidentified Flying Objects* (New York: Bantam Press, 1969).
15. Richard Loye Martin, "Magnesium from the Electrolytic Reduction of Magnesium Oxide" (PhD thesis, University of Arkansas, 1949).
16. Martin, "Magnesium Oxide."
17. Condon and Gillmor, *Scientific Study*.

18. Ibid.
19. Sturrock, "Composition Analysis."
20. Sevik Huseyin and S. Can Kurnaz, "The Effect of Strontium on the Microstructure and Mechanical Properties of mg-6al-0.3mn-1sn," *Journal of Magnesium and Alloys* 2, no. 3 (2014).
21. Qiyang Tan, Yu Yin, Ning Mo, Mingxing Zhang, and Andrej Atrens, "Recent Understanding of the Oxidation and Burning of Magnesium Alloys," *Surface Innovations* 7, no. 2 (2019).
22. Huseyin and Kurnaz, "The Effect of Strontium."
23. M. N. Rao, Razia Saltana, and Sri Harsha Kota, "Electronic Waste," *Solid and Hazardous Waste Management*, 2017.
24. Aytek Yuksel, "What Are Tech Metals and Rare Earth Elements, and How Are They Used?" *Cummins Newsroom*, April 19, 2021 https://www.cummins.com/news/2021/04/19/what-are-tech-metals-and-rare-earth-elements-and-how-are-they-used (accessed November 2021).
25. Paton, Chad Paton, Martin Schiller, and Martin Bizzarro, "Identification of an 84 Sr-Depleted Carrier in Primitive Meteorites and Implications for Thermal Processing in the Solar Protoplanetary Disk," *The Astrophysical Journal Letters* 763, no. 2 (2013).
26. Maria Lugaro, Amanda I. Karakas, Maria Peto, and Emese Plachy, "Do Meteoritic Silicon Carbide Grains Originate from Asymptotic Giant Branch Stars of Super-Solar Metallicity?" *Geochimica et Cosmochimica Acta* 221 (2018).
27. Robert M. Powell, Michael D. Swords, Mark Rodeghier, and Phyllis Budinger, *Isotope Ratios and Chemical Analysis of the 1957 Brazillian Ubatuba Fragment* (Scientific Coalition for UAP Studies, 2021).
28. Lugaro et al., "Do Meteoritic Silicon."
29. Sturrock, "Composition Analysis."
30. Powell et al., *Isotope Ratios*.
31. Sturrock, "Composition Analysis."
32. Powell et al., *Isotope Ratios*.
33. Larry R. Nittler, Timothy J. McCoy, Pamela E. Clark, Mary E. Murphy, Jacob I. Trombka, and Eugene Jarosewich, "Bulk Element Compositions of Meteorites: A Guide for Interpreting Remove-Sensing Geochemical Measurements of Planets and Asteroids," *Antarctic Meteorite Research* 17 (2004): 231–51.
34. Powell et al., *Isotope Ratios*.
35. Ibid.
36. Ibid.

4

The Lonnie Zamora Incident

I have been privileged to be briefed and to know that we have been visited. I do not have firsthand experience in this regard, but I have been on investigating teams, and I have been briefed by insiders who do know. Yes, we have been visited and it appears that our visitors are prepared to help us if we allow them.

—Dr. Edgar Mitchell (Apollo 14 astronaut, sixth man on the moon)[1]

Acknowledgments: I would especially like to thank Rob Mercer and John Greenewald Jr. for their efforts in collecting and preserving the documents referred to in this case.

4.1 Case Description

*T*HE FOLLOWING IS BASED *on the official Project Blue Book file available from the National Archives (identifier number 595466), documents released to John Greenewald Jr. by the FBI via a FOIA request,[2] and documents obtained by Rob Mercer from Lieutenant Carmon Marano, a member of Project Blue Book.[3]*

At 5:45 p.m. on April 24, 1964, Officer Lonnie Zamora, of the Socorro Police Department in New Mexico, sat looking out the window of his white, 1964 Pontiac. Zamora was on patrol about one mile southwest of Socorro in an isolated area. He looked out across the dry landscape of sand and brush. He could hear the distant sound of a car's engine. It grew louder, more aggressive. Until eventually it rushed past at great speed, heading south. Officer Zamora fired up his police car and gave chase. While in pursuit of the speeding car, Zamora glimpsed something strange in the distance to his right. A bright flame-like light in the southwestern sky, about half a mile to a mile away. Then he heard a loud roar from the same direction. Thinking a nearby dynamite shack had exploded, he decided to abandon the chase and investigate.

Zamora turned his Pontiac down a rough gravel road, heading toward the light. The source slowly came into focus. It was a brilliant blue and orange flame, gradually descending. The flame had a thin funnel shape, twice as narrow at the top as at the bottom, and about four times as high as it was wide. There was no smoke. As his car drew nearer, he could make out a disturbance as plumes of dust and sand rose from behind the hill. The noise was also clearer now. It was a sustained roar, not an explosion. The pitch changed from high frequency to low, before eventually stopping. After this, Zamora heard no noise, except that of his car struggling up the steep hill along a rough gravel track. At the top, he traveled westward for around fifteen seconds at a crawl, looking for the source of the light and noise. Glancing south, down the hill, Lonnie Zamora saw something that would change the rest of his life.

There, only six hundred feet away, was a white oval shaped object about twenty feet long. Zamora stopped the car for two seconds to give it his full attention. Two legs slanted outward from the bottom of the craft, holding it about three and a half feet above the sandy ground. The craft appeared to be perfectly smooth. Not a single door, window, or seam was visible. Two figures, dressed in white coveralls, stood close to the craft. Although the figures appeared to have normal human proportions, something was off. They were small, small like children, or possibly small adults. One of the figures turned and noticed Zamora. Startled, it appeared to jump. Thinking a couple of teenagers had overturned their car, Zamora quickly started his car toward them with the idea of helping. On route, he radioed to the sheriff's office, "Socorro 2 to Socorro, possible 10-44 (accident); I'll be 10-6 (busy) out of the car checking the car down in the arroyo." Now at only a hundred feet, Zamora stopped his car and got out to inspect the scene on foot. As he approached, he heard "two or three loud "thumps," like someone possibly hammering or shutting a door or doors hard. These "thumps" were possibly a second or less apart." The two small figures were no longer visible. He could now make out a red symbol printed on the side of the object. Zamora estimated the insignia to be approximately thirty inches high by twenty-four inches wide (0.76 by 0.61 meters). The symbol looked like an upward pointing arrow inside an upside down "U" shape.

Then a sudden roar. Like a jet engine firing up. Only different. This time the sound went from low to high frequency. Light blue and orange flames shot from under the oval object as it slowly rose straight up. Fear gripped

Zamora. He turned and ran away. Slamming into the back fender of his car, his glasses fell to the ground. He left them. Sprinting hard over the rough ground, he had to put distance between him and the object, whatever it was. Reaching the top of the hill he glanced back, shielding his face with his arms. The roar had stopped. The object was moving away to the southwest at a height of only fifteen feet above the ground. The craft now moved in complete silence, with no flame, and at a very high rate of speed. It quickly disappeared over the distant canyons and mountains.

Zamora cautiously walked back to his police car and found his glasses on the floor. He got in and closed the door. He felt a bit safer now. Zamora radioed Socorro Police dispatcher Nep Lopez, demanding, "Look out the window, do you see an object?"

Lopez replied,

"What is it?"

Zamora thought for a minute and said,

"It looks like a balloon."

In response to Zamora's radio calls, Sergeant Chaves of the New Mexico State Police was dispatched to the scene. While waiting for backup, Zamora took out a pen and sketched the red insignia he had seen on the side of the craft. Sergeant Chaves then arrived to find a still very pale, frightened, and sweaty Zamora. Together, they slowly approached where the craft had been. Sergeant Chaves and Zamora discovered clear and deep depressions in the ground as well as burned brush. The brush was charred in several areas but now cold to the touch.

Special Agent D. Arthur Byrnes Jr., of the Federal Bureau of Investigation (FBI), also heard Zamora's radio call. Agent Byrnes left the State Police Office at Socorro at approximately 6:00 p.m. and headed straight to the scene. Agent Byrnes arrived to find Zamora somewhat agitated over his experience but nevertheless clear and perfectly sober. As documented in the official report by the United States Department of Justice,

> Special Agent Byrnes noted four indentations in the rough ground at the "site" of the object described by Officer Zamora. These depressions appeared regular in shape; approximately sixteen by six inches rectangular. Each depression seemed to have been made by an object going into the earth at an angle from a center line. Each depression was approximately two inches deep and pushed some earth to the far side.
>
> Inside the four depressions were three burned patches of clumps of grass. Other clumps of grass in the same area appeared not to be disturbed. One burned area was outside the four depressions.

There were three circular marks in the earth which were smooth, approximately four inches in diameter and penetrated in the sandy earth approximately one-eighth of an inch as if a jar lid had gently been pushed into the sand.

The officers at the scene marked and preserved the larger indentations by placing stones around them. Two of the larger indentations measured by Agent Byrnes can be seen in figure 4.1.

The four indentations defined an irregular four-sided shape on the ground. The longest side measured 14 feet 7.5 inches. The second longest, 13 feet 2.5 inches. The two shortest sides were 11 feet 10.5 inches and 9 feet 7.5 inches, respectively. Three burn marks were found inside the quadrilateral and one outside, some six feet away. The three smaller circular indentations were located within the four-sided shape but away from the center.

Within the first few hours, a total of nine officials saw the markings that Zamora alleged were left by the craft. There was at least one other corroborat-

FIGURE 4.1
A photograph of two of the four indentations. *Credit*: National Archives; Rob Mercer; John Greenewald Jr.

ing witness. An unidentified tourist traveling north on US Route 85 saw what is thought to be the same UFO just before it landed. The manager of Whitting Brothers' Service Station on Route 85 North, Mr. Opel Grinder, said the tourist stopped at the station and claimed an unidentified object flew very low over his car. The tourist also reported that the UFO headed straight for the gully where it would land moments later, and that they had seen a police car heading up the nearby hill, which was presumably Zamora's.

Two days later, on April 26, 1964, a similar incident occurred just sixty miles from Socorro, in a small town called La Madera. In this case, at about 1:00 a.m. in the morning, Orlando Gallegos went out to tend to noisy horses on his father's ranch. Approximately three hundred feet from the house was something "egg-shaped, like a large butane gas tank." The object was twelve to fifteen feet high and "as long as a telephone pole." Gallegos approached to within two hundred feet from the object. It lay stationary on the ground in complete silence and was surrounded by a blue-white flame. It appeared to be made of bright metal and had no visible windows. The witness observed the object for a full minute before the flame went out. At approximately 7:30 p.m. the same day, State Police Captain Martin Vigil investigated the scene. Captain Vigil found a scorched circular area about thirty to forty feet in diameter. At least four rectangular indentations were also found at the scene.

Following the Socorro incident, Lonnie Zamora eventually grew tired of reporters, ufologists, and Air Force personnel asking him questions and frequently turning up at his private residence unannounced. He never sought to gain fame or fortune from the incident. In fact, it could be said that he went the other way, secluding himself from the public, retiring from the police force, and taking a job managing a gas station. Zamora died in Socorro in November 2009, not far from the landing site that would come to define his life.

4.2 Analysis

4.2.1 Weighing a UFO

The craft allegedly left physical evidence when it landed that can be analyzed. As reported by FBI special agent Byrnes the craft left four imprints in the sand. We are going to use this evidence to try to weigh the object that landed in Socorro, New Mexico, that evening in 1964. To the best of my knowledge, this is the first time a UFO has ever been weighed.

How on Earth do you weigh a UFO? Well, it may help to think about dinosaurs. Yes, that's right, dinosaurs. How do we know how much a dinosaur weighed? They're all dead so we cannot just place them on a scale. One method is to look at the gigantic, fossilized footprints they left behind. One can measure the surface area of these prints, and the depth of the depression they left in the ground. Based on these measurements, it is possible to estimate the dinosaur's mass, without ever having to put a T-Rex on a set of weighing scales.

All we need is some simple physics. You see, the pressure a dinosaur's foot exerts on the floor is proportional to the downward force due to its body weight. But that's not all. Pressure is also inversely proportional to the surface area of the bottom of the dinosaur's foot. For example, if you made a T-Rex wear stilettos, then its foot would sink much deeper into sand than if it wore large, flat shoes, even though its body weight is the same. The smaller the contact area between sand and foot the bigger the pressure, and therefore the deeper the print. Put simply, we can estimate the weight of an object based on the surface area and depth of the imprint it leaves in the floor.

According to agent Byrnes the craft left four rectangular indentations in the soft sand, each of them measuring approximately sixteen by six inches. Thus, we know the approximate area of each of the craft's "feet." Crucially, Agent Byrnes also measured the depth of the imprint to be approximately two inches. Therefore, we have everything we need to weigh the Socorro UFO. The result is a total mass of[a]

$$M_{object} = (1789.7 \pm 509.2) \, kg. \tag{4.1}$$

For a large oval-shaped volume that was reported to measure twenty feet across, this is surprisingly light. To put this in perspective, it is about the weight of a large car. This estimated mass can now be used to test two of the leading alternative explanations.

4.2.2 Testing Alternative Hypotheses

One of the most popular attempts at a mundane explanation of the Lonnie Zamora incident is based around NASA. At the time of the incident, NASA was testing a prototype of the Surveyor module, a lunar probe that would eventually go to the moon in 1966 in preparation for the Apollo missions. Records show that the lunar surveyor was being tested out of the nearby Holloman Air Force Base on the very same day the Lonnie Zamora sighting occurred. Furthermore, Surveyor used retrorockets that would have produced a loud roar and brilliant flame, like that reported by Zamora. Lastly, Hughes

---

a. This calculation assumes the object's weight was evenly distributed over the four legs and that the legs were perpendicular to the ground. For a detailed explanation of how this value and its associated error were determined see appendix 4.3.1.

Aircraft Company was charged with building Surveyor, and their employees wore white coveralls with a blue logo. Case closed, right? Not so fast.[4]

There are several reasons to doubt this explanation. The first reason comes from the White Sands Missile Range logbook from April 24, 1964, as shown in figure 4.2a. As we can see from this official log, the Surveyor was only being tested that morning between 07:45 and 11:45, approximately six hours before the Zamora sighting.[5] What's more, the sighting in Socorro was a minimum of twenty kilometers from the very edge of the White Sands Missile Range. It is a stretch to conclude that the time and location of the Surveyor test and the Zamora sighting coincide.

The Surveyor probe, shown in figure 4.2b, does not match the smooth oval shape described by Zamora by any stretch of the imagination. Nor is there even a flat surface large enough to display the red insignia Zamora reported.[6] Furthermore, the lunar module has three circular landing pads, which is clearly at odds with the four rectangular imprints found at the scene. One more thing, Surveyor was only fitted with retrorockets to slow its descent. By design, it was simply incapable of takeoff by itself. It thus needed to be transported everywhere by helicopter, as evidenced in the logbook of figure 4.2a. This is in blatant contradiction with Zamora's statement that the

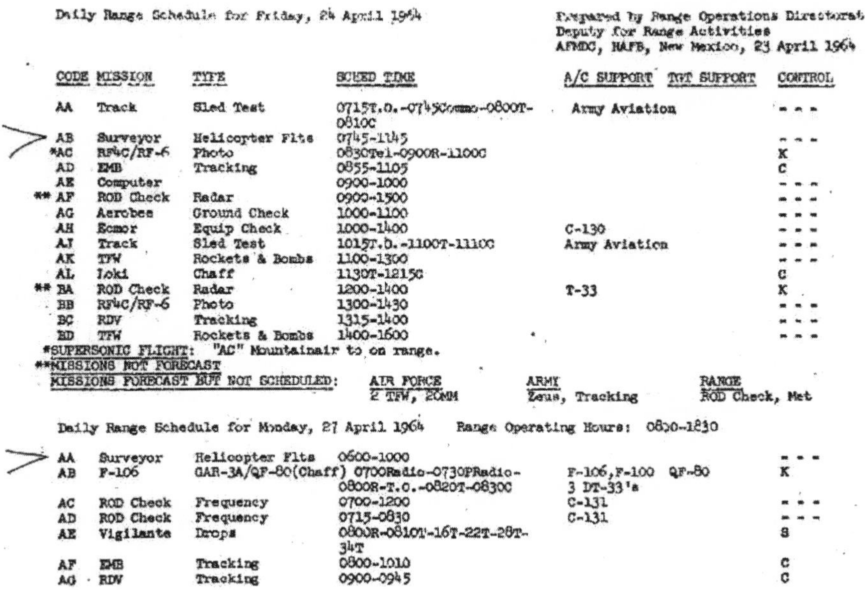

Daily Range Schedule for Friday, 24 April 1964

Prepared by Range Operations Directorat
Deputy for Range Activities
AFMDC, HAFB, New Mexico, 23 April 1964

CODE	MISSION	TYPE	SCHED TIME	A/C SUPPORT	TGT SUPPORT	CONTROL
AA	Track	Sled Test	0715T.O.-0745Commo-0800T-0810C	Army Aviation		- - -
AB	Surveyor	Helicopter Flts	0745-1145			- - -
*AC	RF4C/RF-6	Photo	0830Tel-0900R-1100C			K
AD	EMB	Tracking	0855-1105			C
AE	Computer		0900-1000			- - -
** AF	ROD Check	Radar	0900-1500			- - -
AG	Aerobee	Ground Check	1000-1100			- - -
AH	Ecmor	Equip Check	1000-1400	C-130		- - -
AJ	Track	Sled Test	1015T.O.-1100T-1110C	Army Aviation		- - -
AK	TFW	Rockets & Bombs	1100-1300			- - -
AL	Loki	Chaff	1130T-1215C			C
** BA	ROD Check	Radar	1200-1400	T-33		K
BB	RF4C/RF-6	Photo	1300-1430			- - -
BC	RDV	Tracking	1315-1400			- - -
BD	TFW	Rockets & Bombs	1400-1600			- - -

*SUPERSONIC FLIGHT: "AC" Mountainair to on range.
**MISSIONS NOT FORECAST
MISSIONS FORECAST BUT NOT SCHEDULED: AIR FORCE 2 TFW, 20MM — ARMY Zeus, Tracking — RANGE ROD Check, Met

Daily Range Schedule for Monday, 27 April 1964 Range Operating Hours: 0800-1830

AA	Surveyor	Helicopter Flts	0600-1000			- - -
AB	F-106	GAR-3A/QF-80(Chaff)	0700Radio-0730PRadio-0800R-T.O.-0820T-0830C	F-106,F-100 3 DT-33's	QF-80	K
AC	ROD Check	Frequency	0700-1200	C-131		- - -
AD	ROD Check	Frequency	0715-0830	C-131		- - -
AE	Vigilante	Drops	0800R-0810T-16T-22T-28T-34T			S
AF	EMB	Tracking	0800-1010			C
AG	RDV	Tracking	0900-0945			C

FIGURE 4.2a

(a) White Sands Missile Range logbook for April 24 and April 27, 1964. *Credit*: Capt. James McAndrew

FIGURE 4.2b
(b) A model of Surveyor 1. *Credit*: NASA, public domain

object rose straight up and then moved away at high speed. Not to mention the fact that Zamora does not once report a helicopter, something that would be hard to visually and audially miss.

Now let's analyze the claim that the two figures could have been wearing uniforms from Hughes Aircraft Company. These uniforms were all white except for a blue rectangular logo that measured about nine inches across and four inches high. This blue logo was not mentioned in Zamora's report. Let's try to find out whether Zamora could have seen this logo at the reported distance of six hundred feet (183 meters). The angular resolution of the human eye is proportional to the wavelength of ambient light and inversely related to pupil size. The human eye is most sensitive to light of wavelength 555 nanometers during the day, and 507 nanometers at night. Since the sighting took place in the late afternoon we will take an average of these two values, obtaining a wavelength of 531 nanometers. The pupil diameter depends on how bright it is and can vary from around three millimeters in daylight to around nine millimeters in pitch darkness.[7] Since the sighting was in the late afternoon, we also take an average of these extremes, yielding an estimated pupil diameter of six millimeters. The minimum distance R_{min} Zamora could have resolved can now be calculated, and comes out at

$$R_{min} = (0.0197 \pm 0.0113) \text{ m}. \tag{4.2}$$

This means the smallest object he could have seen is somewhere between 1 and 3 centimeters, or about an inch. Therefore, he should have been able to easily see the nine-inch wide (twenty-three centimeters) blue insignia on their overalls. Yet, this was never reported by Zamora.[b] For more detail on this calculation, see appendix 4.3.2.

Finally, let's examine the Surveyor hypothesis based on the direct physical evidence left by the object that allegedly landed. We know that Surveyor test flights were conducted using a Bell helicopter that would support the craft from its side. One of the lightest Bell helicopters, the Bell 47, has a mass of approximately 858 kilograms.[8] We also need to add the weight of the Surveyor module to this, which had a minimum possible mass of 294.3 kilograms.[9] Thus, the combined weight of the Surveyor module and helicopter would be at least 1,152 kilograms, and probably much more. This value is not unrealistically far from the estimated mass of the Socorro object (M_{object} = (1789.7 ± 509.2) kg). In summary, the Surveyor hypothesis seems unlikely based on the circumstantial evidence. However, based on the direct physical evidence in the form of the ground imprints it cannot be ruled out.

Another popular hypothesis is that it was a hoax perpetrated by students from the nearby New Mexico Institute of Mining and Technology. Part of the reason for this hypothesis comes from the fact that Zamora once worked on campus at New Mexico Tech. During his time on campus, he was apparently not well liked by the students, due to his rigidity and strict adherence to the rules. Plus, the energetics lab at New Mexico Tech had all kinds of pyrotechnics, which could have been used by students to create the sounds and lights Zamora reported. White lab suits would also have been available to students.

---

b. This calculation also verifies that Zamora could have resolved the red symbol he claims was on the side of the craft in even greater detail, since this observation was made at the even closer distance of one hundred feet (thirty meters).

According to the University President, the craft itself consisted of "A candle in a balloon. Not sophisticated."[10] This fits with Zamora's statement to Socorro Police dispatcher Nep Lopez that the object "looks like a balloon."

Let's examine the idea that this sighting could be explained as a hoax by local students using a balloon. As alluded to by the president of New Mexico Tech, the most likely type of balloon is a simple hot air balloon. Whenever you have a substance of lower density surrounded by a substance of higher density it rises upwards due to the so-called buoyant force. For example, an oxygen bubble released deep under water will rapidly rise toward the surface, because oxygen is less dense than the surrounding water. The same thing happens when you put an object in the atmosphere that is less dense than the surrounding air. It rises. This buoyant force can even be used to lift a mass up, such as passengers in a hot air balloon. We can make normal air less dense by heating it, as is done by the burner in a hot air balloon. The hot air inside the balloon becomes less dense than the cold air surrounding it. The result is lift.

Let's assume the balloon is spherical with a diameter equal to twenty feet, which is approximately consistent with Zamora's account. This means the balloon has a volume of 118.6m^3. If we also assume the air outside is twenty degrees Celsius and the air inside the balloon is heated to one hundred degrees Celsius, then we find that 1m^3 of this hot air can lift a mass of 0.259 kilograms at sea level. Therefore, the maximum mass that can be lifted by a hot air balloon of this size is

$$M_{max}(hot\ air) = 30.72\ kg. \qquad (4.3)$$

This is quite a small mass, which is why hot air balloons need to be so big. The estimated mass of the object that landed in Socorro, based on the physical imprints it left behind, is therefore a factor of about fifty-eight greater than the maximum possible mass a hot air balloon of the same size could have. The object was therefore unlikely to have been a hot air balloon.

The balloon could also have been filled with a gas other than air, for example hydrogen or helium. These gasses are among the most common gasses used in balloons with lift and would have been readily available to New Mexico Tech students. Since hydrogen and helium are less dense than air, these gasses will also create lift in the atmosphere. We know that 1m^3 of hydrogen in air at sea level can lift a mass of 1.202 kg. Therefore, the maximum mass that can be lifted by a volume of 118.6m^3 of hydrogen is

$$M_{max}(hydrogen) = 142.6\ kg. \qquad (4.4)$$

Likewise, given that 1m³ of helium can lift 1.114 kg in air, this means that 118.6m³ of helium can lift a maximum mass of

$$M_{max}(helium) = 132.1 \ kg. \tag{4.5}$$

These masses are about thirteen times smaller than the mass of the Socorro object based on the physical imprints it left behind. This analysis also makes it unlikely that the UFO Zamora witnessed was a hydrogen or helium balloon.

4.2.3 Weighing an Alien

There was one more piece of physical evidence found at the scene that we haven't mentioned yet—possible footprints. Within the four rectangular indentations thought to be left by the craft were a few small depressions in the soft sand described as footprints and tracks in the documentation. These indentations were approximately trapezoidal in shape, measuring roughly 4.5 inches by 4 inches (11.4 by 10.2 centimeters). The impressions were only about 0.5 inches deep (1.3 centimeters). Applying the same method, we used to weigh the UFO, we can also weigh whatever left these prints. The result, assuming the weight is evenly distributed over two legs as implied by Zamora's account of a small human-looking being, is

$$M_{2legs} = (41.9 \pm 16.8) \ kg. \tag{4.6}$$

Did we just weigh an extraterrestrial? Possibly, but these prints could also have been made by a lot of other things. If we assume for a minute that the prints were made by a bipedal human-like entity, then this weight is unusually low. Using a standardized conversion chart between height and weight implies the entity would have a height of somewhere around four feet ten inches (148 centimeters).[11] This is the height of a typical eleven-year-old boy.[12] This estimate is therefore remarkably consistent with Zamora's description of the two figures resembling children or small adults. However, several four-legged predators are known to live in New Mexico such as the mountain lion, black bear, bobcat, and coyote. Instead, assuming the prints were made by a weight evenly distributed over four legs, instead of two, gives a larger mass estimate of

$$M_{4legs} = (83.8 \pm 33.6) \ kg. \tag{4.7}$$

This mass is too big to be a bobcat or coyote, but it is not inconsistent with the mass of an adult mountain lion or black bear. However, for a trained Police officer to mistake a mountain lion or black bear for teenagers dressed in white coveralls at six hundred feet somewhat strains credulity.

4.2.4 Evaluation of Evidence

Here we evaluate and summarize the evidence related to the Lonnie Zamora incident. The Lonnie Zamora incident is widely regarded as one of the best documented UFO cases in history. The FBI investigated the scene and concluded it could not be explained. The US government also investigated the incident as part of Project Blue Book and concluded that the UFO sighting and landing could not be explained.

Eyewitness Testimony

The reliability of this case is in large part due to Lonnie Zamora himself. The FBI assessed him in depth as a witness and concluded that Officer Lonnie Zamora "is well regarded as a sober, industrious, and conscientious officer and not given to fantasy." In another document by the United States Air Force (USAF), it was stated that "All persons interviewing Zamora are impressed with his sincerity and are of the belief that if a hoax has been perpetrated Lonnie Zamora is definitely not a part to it." In his official statement on the incident, Lonnie Zamora said that he was in good health and that his last alcoholic drink was just "two or three beers–over a month ago."

In all, nine witnesses observed the physical evidence within hours of the incident, including Officer Zamora, Sergeant Chavez, Captain Holder, FBI Agent Byrnes, and a team of Air Force investigators from Wright-Patterson Air Force Base. An anonymous tourist at a nearby service station on Route 85 North also witnessed presumably the same UFO before it landed. The second, possibly related incident in nearby La Madera, was investigated by Captain Vigil. The quantity and quality of the eyewitness's testimony following the alleged landing is excellent. However, the actual UFO itself was only witnessed by Officer Zamora and so we downgrade this category of evidence slightly.

Single and Multiple Sensor Detection

Several photographs were taken of the landing site, including the indentations left by the craft and the burned patches of grass. A USAF document disclosed that a Major Conner and Sergeant Moody used a Geiger counter to check the area of the UFO sighting for radiation. The levels were found to be normal. Captain Holder asked radar operators at White Sands and Alamogordo if there was any unusual activity during the time of the incident.

The reply was negative. Therefore, despite several attempts to collect sensor data the only positive results come from photographs of the indentations and burn marks. The evidence from single and multiple sensor detection is therefore considered lacking in quantity, quality, and reliability. However, the sources of this limited data, namely the USAF and FBI, are considered reliable.

Physical Evidence

There are four pieces of physical evidence in this case. The larger rectangular indentations, the smaller trapezoid indentations, the shallow circular indentations, and the burn marks in the soil and shrubs. Soil samples from the site were tested via spectrographic analysis but no foreign material was reported. Strangely, this analysis did not detect any chemical propellant in the burn marks, which one would expect from a surveyor-type retrorocket. The quantity of physical evidence is therefore good, but not excellent.

Regarding the quality, consistency, and source of the physical evidence, measurements of the indentations, and burn marks were carefully measured and recorded via detailed sketches and photographs by both FBI Agent Byrnes and Captain Holder, in addition to the USAF investigation team. All measurements are at least approximately equivalent and therefore considered largely consistent. The source of the physical evidence is considered reliable as it comes mainly from the FBI, the New Mexico Police Department, and the USAF. In a statement by the USAF, dated May 28, 1964, it is stated that "Information obtained during this investigation revealed that the sighting was legitimate and there was no indication that a hoax was being perpetrated."

Summary

In this analysis, it is possible that we weighed a UFO and an extraterrestrial for the first time. However, it is also possible that we did not. The estimated weight of the UFO indicates it is extremely unlikely to be a hot air balloon, a hydrogen balloon, or a helium balloon. But it could be many other things. For example, circumstantial evidence makes it unlikely that the object Zamora witnessed was the Surveyor module, however the physical evidence cannot rule this out. Also, the possible footprints do yield an unusually small mass for an adult human, which may lead one to speculate the footprints were made by the typical small and thin depiction of an alien. Yet, we must caution against such speculation, since the prints are also consistent with an adult mountain lion or black bear.

The Lonnie Zamora incident is often described as one of the best documented UFO cases. I would not disagree with this. It is a compelling case based on the eyewitnesses and the physical evidence, yet it is lacking in single and multiple sensor data. The evaluation of evidence in the Lonnie Zamora case is summarized in table 4.1.

TABLE 4.1
A table summarizing the evaluation of evidence in this case.

Evidence type	Quantity	Quality	Consistency	Source	Total	Weighted total
Eyewitnesses	2	3	3	3	11	11
Single sensor data	1	1	1	3	6	12
Multiple sensor data	0	0	0	0	0	0
Physical evidence	2	3	3	3	11	44

Final verdict: 67/120 (56 percent)

4.3 Appendix

4.3.1 How to Weigh a UFO

In physics, pressure P is defined as the perpendicular force F applied to a surface per unit area A. Mathematically, we can write this as

$$P = \frac{F}{A}. \tag{4.8}$$

We assume that the pressure exerted by an object on sand is directly proportional to the depth of the resulting imprint. Mathematically, we can write this statement as

$$P = kD, \tag{4.9}$$

where P is the pressure, k is a constant of proportionality and D is the depth of the indentation. We have empirically tested the validity of this assumption up to pressures of approximately 25,000 Pa. This experiment involved placing eight different blocks of known mass and cross-sectional area on a level layer of sand. Since the force due to the block's weight is known, as well as its cross-sectional area, the downward pressure exerted on the sand can be calculated via equation 4.8. The depth of the depression each weight made in the sand was then carefully measured. The results are displayed in figure 4.3, which verifies an approximately linear relationship between pressure and imprint depth up to about 25,000 Pa.[c]

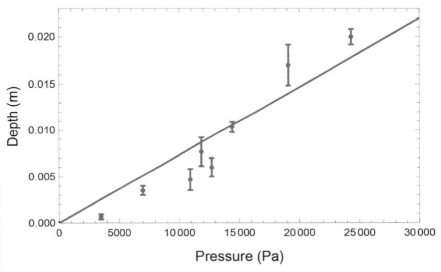

FIGURE 4.3
Applied pressure against indentation depth in sand.

Applying equation 4.9 to the specific cases of the block and the unknown object, taking their ratio and rearranging gives

$$P_{object} = \frac{D_{object}}{D_{block}} P_{block}. \qquad (4.10)$$

Using equation 4.8 and F = Mg, where g = 9.81 ms^{-2} is the acceleration on Earth due to gravity, yields

$$M_{object} = \frac{A_{object} D_{object} P_{block}}{D_{block} g}. \qquad (4.11)$$

Reading off values for P_{block} and D_{block} from figure 4.3, and by obtaining measurements of the area A_{object} and depth D_{object} of the print left by the object, one can determine the mass of the unknown object via equation 4.11.

---

c. The proportionality defined by equation 4.9 is assumed to also apply for much heavier objects, or greater pressures, as assumed elsewhere in the literature. For example, this proportionality is sometimes assumed when estimating a dinosaur's weight based on the depth of its footprint.

Alternatively, we could write equation 4.11 more explicitly as

$$M_{object} = \frac{A_{object} D_{object} M_{block}}{A_{block} D_{block}}. \quad (4.12)$$

The value given in equation 4.1 was determined using equation 4.12, for specific measured values M_{block} = 4.1534 kg, A_{block} = 0.002827 m^2, D_{block} = 0.01033 m. The uncertainty in the value of M_{object} was estimated by applying a propagation of error analysis to equation 4.12, with all length measurements of the impression made by the object having uncertainty ±0.0127 m and for the footprints ±0.00318 m. Due to the accuracy of the weighing scales used the uncertainty in the mass of the test block is taken to be ΔM_{block} ± 0.05 g.

4.3.2 Resolution of the Human Eye

The angular resolution θ of a circular aperture, for example the pupil of the human eye, is given by

$$\theta = 1.22 \frac{\lambda}{D}, \quad (4.13)$$

where λ is the wavelength of light and D is the diameter of the aperture, in this case the pupil. The minimum resolvable distance R is therefore given by the arc length equation $R = d\theta$, where d is the distance between the observer and the object being resolved. Hence,

$$R = d\theta = 1.22 \frac{d\lambda}{D}. \quad (4.14)$$

The uncertainty in the value of R can then be estimated by applying a propagation of error analysis to equation 4.14, with error estimates Δd = ±50 m, $\Delta \lambda$ = ±25 nm, and ΔD = ±3 mm.

4.3.3 The Physics of Balloons

The buoyant force F_b generated by a balloon of volume filled with a gas of density ρ_{gas} suspended in air of density ρ_{air} is given by

$$F_b = mg = (\rho_{air} - \rho_{gas})gV. \quad (4.15)$$

Rearranging equation 4.15 gives

$$\frac{m}{V} = (\rho_{air} - \rho_{gas}). \qquad (4.16)$$

Plugging in values for the density of air and the respective gas gives (1.292 − 0.090) = 1.202 kgm^{-3} for Hydrogen, and (1.292 − 0.178) = 1.114 kgm^{-3} for Helium. The same method was applied in the study of the hot air balloon.

In our calculations we have assumed the balloon is spherical in shape and thus its volume is given by the equation

$$V_{sphere} = \frac{4}{3}\pi r^3, \qquad (4.17)$$

where r = 10 ft = 3.048 m is the radius of the spherical balloon. However, Zamora described the object as oval-shaped, but this would only lead to a smaller volume of gas and therefore a smaller maximum mass.

4.4 Notes

1. Tim Miejan,. *The Way of the Explorer: An Interview with Dr. Edgar Mitchell*, July 1, 2008 (accessed May 10, 2022, https://www.edgemagazine.net/2008/07/edgar-mitchell/).

2. John Greenewald, *theblackvault*, n.d. (accessed March 10, 2022, https://documents2.theblackvault.com/documents/fbifiles/paranormal/FBI-UFO-Socorro-fbi1.pdf).

3. Rob Mercer, *From the Desks of Project Blue Book*, n.d. (accessed March 10, 2022, https://www.theblackvault.com/casefiles/desks-project-blue-book/).

4. Brian Dunning, *Lonnie Zamora and the Socorro UFO*, August 1, 2017 (accessed March 2, 2022, https://skeptoid.com/episodes/4582).

5. David E. Thomas, *New Mexicans for Science and Reason; The Socorro, NM UFO*, n.d. (accessed May 29, 2022, http://www.nmsr.org/socorro.htm).

6. Dunning, *Lonnie Zamora*.

7. *Resolution of Human Eye*, 2017 (accessed March 14, 2022, https://www.wikilectures.eu/w/Resolution_of_human_eye).

8. *The Bell 47 Light Helicopter Project, USA* (accessed March 10, 2022, https://www.aerospace-technology.com/projects/bell47lighthelicopte/).

9. *Surveyor 1* (accessed March 2022, https://nssdc.gsfc.nasa.gov/nmc/spacecraft/display.action?id=1966-045A).

10. Dunning, *Lonnie Zamora*.

11. *Height and Weight Chart* (accessed March 10, 2022, https://www.forumhealthcentre.nhs.uk/your-health/height-weight-chart).

12. *halls.md* (accessed March 10, 2022, https://halls.md/chart-boys-height-w/).

5

The Aguadilla Object

At the time that I saw it, I said there was something out in front of me or outside the spacecraft that I couldn't identify, and I never have been able to identify it, and I don't think anybody ever will.

—General James McDivitt (NASA astronaut)[1]

Acknowledgments: I am especially indebted to the Scientific Coalition of UFOlogy (SCU) for their excellent analysis of this case and for acquiring some of the raw data.[2]

5.1 Case Description

AT 8:58 P.M. ON APRIL 25, 2013, a radar installation located ninety-two miles east-southeast of Pico Del Este, Puerto Rico, began blipping. It was picking up radar reflections from a single object located two to three miles to the northwest of the Rafael Hernández International Airport. What was odd about these radar returns was that they had no transponder code. Military, police, and commercial aircraft transmit a unique signal, a transponder code, that identifies them to radar installations. This is crucial for air traffic safety and law enforcement. At least forty-six radar pulses bounced back from this object over a sixteen-minute period. Not one had an identifying code. What's more, the object was moving in a highly unusual way, rapidly changing speed, and direction.[3] At 9:14 p.m., the radar installation lost track of the unknown object.

Just two minutes later, at 9:16 p.m., a DHC-8 turboprop aircraft operated by US Customs and Border Protection took off on a routine mission from the Rafael Hernández Airport, initially heading east. Onboard were the pilot, co-pilot, and two instrument operators. As the plane made a course correction to the northwest the crew spotted a pink-red light over the ocean. Without knowing it, the crew of the US Border Protection aircraft had just visually corroborated the unusual radar returns two to three miles to the northwest. The object appeared to be heading in the direction of the airport. The pilot relayed the sighting to the airport control tower, who also confirmed visual contact.

At 9:20 p.m., Flight FX58 sat on the runway of the Rafael Hernández Airport. The crew were finalizing their pre-flight safety checks when an unidentified flying object suddenly and dangerously flew at low altitude right across the runway. The object, whatever it was, once again could not be identified via a transponder code and the control tower received no radio communication from the object. With a clear and present flight safety risk, the departure of flight FX58 bound for Memphis was delayed until further notice.

The Border Protection aircraft continued to monitor the unknown object. The radar onboard the DHC-8 did not detect the UFO, possibly because it was specifically designed to detect seacraft not aircraft. The instrument operator then switched on the thermal imaging video system—bingo. The object appeared bright and clear on the onboard display. Its black signature on the infrared video implied it was hotter than the surroundings. The temperature of the object can be estimated from the thermal video, with the center being approximately 105 degrees Fahrenheit (41 Celsius). The object was tracked by the thermal imaging system for a total of two minutes and fifty-six seconds. This video contains a wealth of information that can be carefully analyzed for clues.

The video shows a small object that is hotter than its surroundings flying low over the land near Aguadilla, heading out towards the Atlantic Ocean. After getting about four hundred meters out to sea the object enters the water. It continues underwater at a high rate of speed, before emerging from the waves. Moments later the most bizarre aspect of the whole incident occurs. The object can be seen to divide in two. Two distinct objects then continue to fly over the Atlantic, gradually moving apart.

5.2 Analysis

5.2.1 Radar Data

The radar installation known as QJQ is located ninety-two miles east-southeast of Pico Del Este, Puerto Rico, at an elevation of 1042 meters above sea level. It is an FPS-20E L-band radar that operates at a frequency of 1280

The Aguadilla Object

to 1350 MHz and has a maximum range of two hundred nautical miles. The radar completes one full sweep every twelve seconds.[4] The QJQ radar detected at least forty-six primary radar returns (primary meaning there was no transponder code) from the unknown object between 8:58 p.m. and 9:14 p.m., local time. Based on the actual radar returns, the path the UFO took over the northwest coast of Puerto Rico during this time is shown in figure 5.1. Note that only the latitude and longitude of the UFO are shown in figure 5.1, not the altitude of the craft above sea level.

The UFO was first picked up by the radar in the upper left of the picture. It begins heading southwest over the course of a few radar sweeps before abruptly changing course and appearing to jump approximately nine kilometers almost due east. This *jump* is labeled J1 in figure 5.1. The path then becomes extremely erratic, performing what looks like a search pattern that slowly tends toward the southwest. We point out that such an erratic flight path would be near impossible for a commercial airplane. The UFO then makes a second large *jump* almost due south, labeled J2 in figure 5.1, before vanishing from radar.

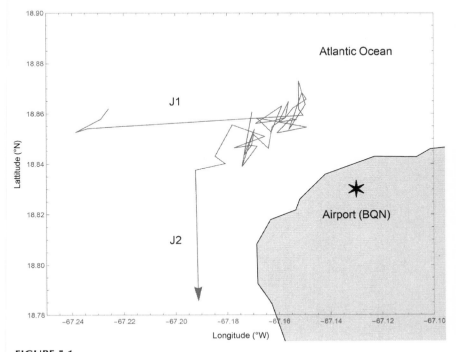

FIGURE 5.1
The flight path of the Aguadilla object based on primary radar returns to the northwest of the Rafael Hernandez International Airport (BQN).

We can estimate the speed with which the object traversed J1 and J2 to try to narrow down what this object could have been. The distance across J1 can be determined via the latitude and longitude of the beginning and end points (see appendix 5.3.1 for details), and comes out at 8812 meters. The data from the QJQ radar tells us that the time between these radar returns was 12.059 seconds.[5] Likewise, J2 equates to a distance of 5,388 meters in a time of 12.137 seconds. The speeds V_{J1} and V_{J2} of the UFO over these two *jumps* are therefore found to be

$$V_{J1} = (730.7 \pm 30.3) \ ms^{-1}, \qquad V_{J2} = (443.9 \pm 18.4) \ ms^{-1}. \qquad (5.1)$$

These speeds are staggering. They rule out all helicopters and weather balloons, and nearly all fighter jets. But what makes them bizarre is that V_{J1} is 2.1 times the speed of sound, and V_{J2} is 1.3 times the speed of sound, yet not a single sonic boom was reported by any witness.[a]

5.2.2 Thermal Imaging Data

Just minutes later, presumably the same UFO was picked up by the thermal imaging system onboard the DHC-8 Turboprop aircraft operated by US Customs and Border Protection. The thermal imaging system is a state-of-the-art Wescam MX-15D with a sensor sensitivity between three and five microns.[6] It was tracked by this sensor for close to three minutes. Four important frames from this video are shown in figure 5.2.[7]

Figure 5.2a shows the UFO as it reaches the Atlantic coastline. We choose to start our analysis with this frame because at this point the object drops below an altitude of five miles for the first time, which reduces the difference between target and ground coordinates to an acceptable level. We define point one via the latitude, longitude, altitude, and time displayed in figure 5.2a (see table 5.1). Figure 5.2b shows the object approximately nine seconds later just before it enters the water. These coordinates define point two, as summarized in table 5.1. Using the Heaversine formula, the distance between points one and two is found to be $d_{12} = 361.3 \ m$ (see appendix 5.3.1). Therefore, the speed of the object between these two points is estimated to be

$$V_{12} = (40.1 \pm 3.1) \ ms^{-1}. \qquad (5.2)$$

---

a. These calculations assume the start and end points of each jump were at the same altitude. If the altitudes were different this would only lead to a greater jump velocity. Thus, V_{J1} and V_{J2} provide a lower bound on the objects speed, which could have been significantly greater. For details on how the uncertainty in these values was determined, see appendix 5.3.2.

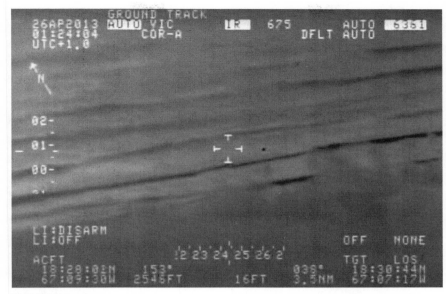

FIGURE 5.2a
Four key frames from the infrared thermal imaging video of the Aguadilla object; (a) The object reaches the northwest coast of Puerto Rico; (b) It enters the Atlantic Ocean; (c) It exits the Atlantic Ocean; (d) It appears to divide into two identical objects. *Credit*: DHS/CPB, video hosted by John Greenewald Jr., theblackvault.com

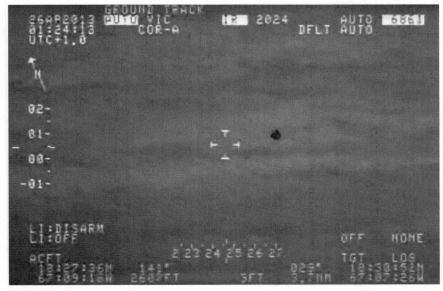

FIGURE 5.2b
Credit: DHS/CPB, video hosted by John Greenewald Jr., theblackvault.com

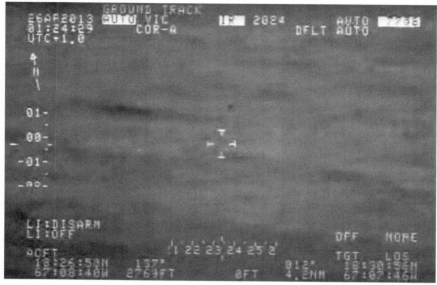

FIGURE 5.2c
Credit: DHS/CPB, video hosted by John Greenewald Jr., theblackvault.com

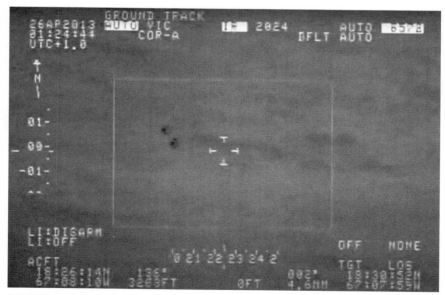

FIGURE 5.2d
Credit: DHS/CPB, video hosted by John Greenewald Jr., theblackvault.com

TABLE 5.1
A table of the latitude, longitude, altitude, and UTC of the four key frames in figure 5.2.

Point (Frame)	Latitude (N)	Longitude (W)	Altitude (m)	UTC
1 (a)	18:30:44	67:07:17	4.9	01:24:04
2 (b)	18:30:52	67:07:26	0.9	01:24:13
3 (c)	18:30:56	67:07:46	0	01:24:29
4 (d)	18:30:52	67:07:59	0	01:24:44

Figure 5.2c captures the object just after it exits the Atlantic Ocean. These coordinates define point three. By applying the Heaversine formula again we can find the distance the object traveled underwater based on its latitude and longitude at points two and three, with the result $d_{23} = 598.7$ m. According to the thermal imaging video it took the object sixteen seconds to cover this distance. Therefore, the underwater speed of the object between points two and three is given by

$$V_{23} = (37.4 \pm 1.7)\ ms^{-1}. \tag{5.3}$$

The speed given by equation 5.2 tells us something interesting. The weather report for the evening of April 25, 2013, tells us that the wind speed was approximately 4.2 ms^{-1} at ground level.[8] At an altitude of five meters the wind speed could have reached as high as 10 ms^{-1}.[9] Even this upper estimate is still about a factor of four smaller than the speed of the object. Moreover, the direction of the wind differs from the object's bearing. It is therefore highly unlikely that the UFO is a purely wind-powered craft such as a balloon. The UFO's airspeed also implies it likely had some form of self-propulsion, yet no hot exhaust signature can be seen in the thermal imaging video.

The underwater speed given by equation 5.3 is even more telling. First, the object hits the water at about 40 ms^{-1} (90 mph) but it does not significantly slow down. In fact, it maintains almost the same speed (37.4 ms^{-1} (84 mph)). It's as if the object does not even notice the water at all. Normally, an object hitting water at this speed and angle would either bounce straight off the surface, like a stone skipping over a lake, or encounter a rapid deceleration if it entered the water due to the immense drag force. Neither of these things happened. The UFO just continued, seemingly oblivious to the fact that it had just entered water. An underwater speed of 37.4 ms^{-1} is also remarkable. The world's fastest submarine can only go about 23 ms^{-1},[10] and the fastest fish in the ocean, the sailfish, can only go about 30.4 ms^{-1}.[11] The data suggests the object was capable of self-propulsion in both air and water, something that is not easily done.

Figure 5.2d shows an image of the object(s) shortly after exiting from the water. The thermal imaging video shows that, starting at approximately UTC 01:24:41, the initially single object begins to expand in size. In under a second, the diameter of the thermal image approximately doubles. The enlarged object now emits bimodal heat signatures, meaning there are now two central sources of heat within the same single object. Finally, the single object appears to physically divide in two. However, each half appears to be about the same size as the single original object, although this is admittedly hard to confirm with any precision due to limitations of the video. A careful frame-by-frame analysis conducted in the robust article of the SCU[12] finds no evidence of any pre-existing second object that joined the first, nor is there any indication that the second object is an infrared reflection of the first. This whole sequence of events is astonishing and reminiscent of mitosis in biology, where a parent cell divides and gives rise to two genetically identical daughter cells.

5.2.3 Evaluation of Evidence

Eyewitness Testimony

Eyewitnesses in this case include the four crew members onboard the DHC-8 turboprop aircraft operated by US Customs and Border Protection (CBP). These crew members remain anonymous due to a non-disclosure agreement.[13] However, we know the rank of the crew members to be a captain, a copilot, and two instrument operators. Since these witnesses were employed by the CBP division of US Homeland Security they are considered reliable. In addition, there were an unknown number of Federal Aviation Authority (FAA) tower personnel that witnessed the incident. Presumably, personnel onboard Flight FX58 also witnessed the unknown pass over the runway of the Rafael Hernández Airport, although this cannot be confirmed.

The sun set at 6:48 p.m. on the day of the incident, and so the visual eyewitness observations were made in the dark. However, that night there was a full moon that rose above the horizon at 6:53 p.m., which would have provided a significant amount of light by the time the incident began at around 9:00 p.m. At the time of the incident there were only scattered clouds and visibility was approximately ten miles.[14]

Therefore, there are at least four reliable and trained eyewitnesses in this case that all consistently report a pink-reddish light in the same direction. Viewing conditions were good enough to enable accurate observations. A particular downside to the eyewitness testimony in this case is their anonymity. Since I cannot study the eyewitness's reliability on an individual basis, only based on their rank and position, the source of these observations is downgraded slightly.

Single and Multiple Sensor Data

A Freedom of Information Act request was made to the USAF Eighty-Fourth RADES group to obtain all FAA radar data during the incident. The request was granted and useful radar data from two independent sites was provided to the authors of the SCU report.[15] The primary FAA radar site was the QJQ installation already mentioned. However, data was also obtained from the so-called SJU radar site near San Juan, Puerto Rico.[16]

An original copy of the thermal imaging video in AVI format was provided by an anonymous source to members of the Scientific Coalition for UFOlogy (SCU) on October 21, 2013. The radar data from SJU verifies that the times and locations displayed on the thermal imaging video match the movements of an aircraft with a government transponder code. This helps to validate the authenticity of the thermal imaging video taken from the CBP aircraft, and all but eliminates the possibility of a hoax. The radar data provided by the USAF from the QJQ site is the source of the forty-six primary radar returns of the unknown object to the northwest of the BQN airport (see figure 5.1). This incident therefore involves multiple corroborating sensor data either directly from a reliable source (in the case of the radar tracks), or that can be verified by a reliable source (in the case of the thermal imaging video).

Physical Evidence

The only glaring weakness in this case is the complete lack of physical evidence.

Summary

The Aguadilla case contains some truly remarkable aspects. For one, an object that to this day remains unidentified was visually observed by at least four trained eyewitnesses, tracked on radar for sixteen minutes, and recorded using an infrared thermal imaging system for three minutes. Another remarkable, almost physics defying, aspect of this case is the fact that the object appeared to go faster than the speed of sound on two separate occasions but produced no sonic boom. This feature seems to be a signature of UFOs as it also appeared in the Japan Airlines Flight 1628 case. Moreover, the object showed essentially no change in speed when entering the ocean despite the significant drag force that should have been present at that speed. Finally, to top of all the weirdness, the object then comes out of the ocean and appears to divide in two.

A summary of the evidence in the case of the Aguadilla object is provided in table 5.2.

TABLE 5.2
A table summarizing the evaluation of evidence in this case.

Evidence type	Quantity	Quality	Consistency	Source	Total	Weighted total
Eyewitnesses	3	3	3	2	11	11
Single sensor data	3	3	3	3	12	24
Multiple sensor data	3	3	3	3	12	36
Physical evidence	0	0	0	0	0	0

Final verdict: 71/120 (59 percent)

5.3 Appendix

5.3.1 Haversine Formula

The distance d between two points with latitudes l_1 and l_2, and longitudes λ_1 and λ_2, is given by the so-called Haversine formula

$$d = 2r \, sin^{-1}\left(\sqrt{sin^2\left(\frac{l_2 - l_1}{2}\right) + \cos(l_1)\cos(l_2)\, sin^2\left(\frac{\lambda_2 - \lambda_1}{2}\right)}\right), \quad (5.4)$$

where r is the radius of the Earth.

5.3.2 Error Estimates

Error estimates quoted in equations 5.2 and 5.3 are obtained via a propagation of error analysis with an uncertainty in time measurements of $\Delta t = \pm 0.5\ s$, since the recorded UTC is only given to the nearest whole second. The uncertainty in distance measurements of $\Delta d = \pm 20\ m$ is estimated based on the Heaversine formula introducing an uncertainty of 0.5 percent due to the Earth being an oblate spheroid not a perfect sphere, in addition to error estimates from crosshair alignment, altitude corrections, and the latitude and longitude only be known to within 0.5 seconds of a degree.

Error estimates quoted in equation 5.1 are obtained via a propagation of error analysis with an uncertainty in time measurements of $\Delta t = \pm 0.5\ s$, which comes from the uncertainty in the time between each radar sweep, and an uncertainty in distance measurements of $\Delta d = \pm 20\ m$ as before.

5.4 Notes

1. Drew Techner, *UFOs Are Real (1979)*, n.d. (accessed May 22, 2022, https://www.youtube.com/watch?v=rpZ7rk_XTrg&t=1151s).
2. Ibid.
3. Scientific Coalition of UFOlogy (SCU), "2013 Aguadilla Puerto Rico," *explorescu.org*, n.d. (accessed May 10, 2022, http://docs.wixstatic.com/ugd/299316_9a12b53f67554a008c32d48eff9be5cd.pdf).
4. SCU, "2013 Aguadilla."
5. Ibid.
6. Ibid.
7. US Customs and Border Protection (CBP), *Anonymous Letter Confirms Aguadilla, Puerto Rico, Coast Guard UFO Video*. April 4, 2015 (accessed April 2022, https://www.theblackvault.com/casefiles/anonymous-letter-confirms-aguadilla-puerto-rico-coast-guard-ufo-video/).
8. "timeanddate," n.d. (accessed May 22, 2022, https://www.timeanddate.com/weather/@4562512/historic?month=4&year=2013).
9. SCU, "2013 Aguadilla."
10. "Underwater Speed Record," *Wikipedia*, n.d. (accessed May 22, 2022, https://en.wikipedia.org/wiki/Underwater_speed_record).
11. "Sailfish," *National Geographic*, n.d. (accessed May 22, 2022, https://www.nationalgeographic.com/animals/fish/facts/sailfish).
12. SCU, "2013 Aguadilla."
13. SCU, "2013 Aguadilla."
14. Ibid.
15. Ibid.
16. Ibid.

II
THE BIGGER PICTURE

6
Where Are All the UFOs?

> I think that we are not alone in the Universe. I believe that someone or something of extraterrestrial origin has visited Earth.
>
> —Colonel Victor Afanasyev (Russian cosmonaut)[1]

6.1 A UFO Toolkit

SUPPOSE FOR A MINUTE that UFOs are extraterrestrial technology, then why would they want to visit Earth in the first place? What does Earth have that they would travel tens, hundreds, or even thousands of light-years to get? Over the years, many ideas have been put forward, some more plausible than others. In this chapter, we investigate several the most popular proposals, including the possible connection between UFO sightings and: the military (including nuclear weapons), water, the environment, earthquakes, and rather strangely blood type.

Instead of focusing on specific cases, we will now take a much broader perspective, analyzing UFO sightings on a national, international, and global scale. In this chapter, we will mainly be concerned with *where* UFO sightings occur most frequently, which will hopefully give us a clue as to *why*. To this end, we will typically analyze the number of UFO sightings within different localized regions of the world, such as within the fifty-one states of the United States, or various countries around the world. After this, in chapter 7, our attention will then turn to *when* most UFO sightings occur, and why. The hope is that by pinpointing when and where UFOs are most frequently sighted, we may get a better understanding of their possible motivations, assuming they have a non-prosaic explanation. Alternatively, if all UFO sightings have a prosaic explanation, we may get an insight into what natural phenomena might be responsible.

To reliably investigate these possible connections, we need a toolkit. Our primary tool will be correlation. *What is correlation you ask?* Well, let's say it's the early 1940s and people are just starting to suspect that smoking may be bad for you. Disturbingly, before this time, cigarettes were generally promoted as having health benefits, with advertisements even encouraging the youth and pregnant moms to take up smoking. The behemoth cigarette companies would not be pleased if you make unfounded accusations to the contrary in public. You need proof or at least strong evidence. So, you find out how many cigarettes are sold per person in a particular country and the number of lung cancer cases per person. But this is just one data point, it tells you nothing. You need to repeat this data collection process for many other countries and see if a trend starts to emerge. *Are there more lung cancer cases in countries that consume more cigarettes?* After sampling say ten countries your data collection may end up looking something like figure 6.1.[a] But how can we know from this graph whether there is a real correlation between cigarette consumption and cancer rates?[2] Well, the trend line does go up, indicating there is some kind of association. Namely, as you move to higher values on the horizontal axis (cigarettes consumed) you tend to see higher corresponding values on the vertical axis (cancer rates). But what if this small number of data points just appeared to give a correlation, purely by chance? Will the correlation start to fade with more data? Alternatively, if the correlation does prove to be real, how strong is it exactly? Should tobacco smoking become illegal immediately, or could it be phased out slowly over several years? To answer such questions, it is useful to have quantitative measures of correlation.

Enter Pearson's correlation coefficient, or simply the r-value, a number that quantifies how strongly one variable is correlated with another. An r-value is a number between negative one and positive one. A positive r-value, which is a value between zero and one, indicates a positive association between the variables. Meaning that as one variable increases the other tends to also increase. Conversely, a negative r-value, that is a value between zero and minus one, indicates a negative association. Meaning that as one variable increases the other tends to decrease. The strength and direction of a given correlation is defined according to its r-value, as shown in table 6.1.

---

a. This data is fictitious and is purely for explanatory purposes. However, in 1950, five real studies all confirmed that smokers of cigarettes were far more likely to contract lung cancer than non-smokers.

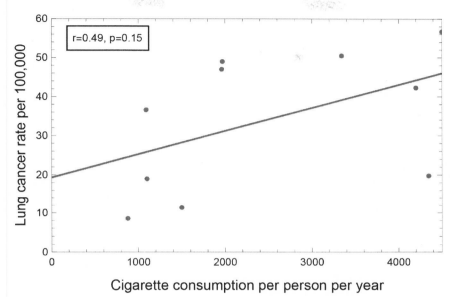

FIGURE 6.1
An illustrative example of the association between cigarette consumption per person per year and lung cancer rates per 100,000 people for ten different countries.

Note that if the r-value is very close to zero (within 0.1) then it is defined as a negligible correlation, meaning any possible correlation is too small to count and is discarded.[b]

TABLE 6.1
A table showing the nomenclature for correlation strength in this work based on its r-value.

r-value	Correlation
0.5 to 1.0 (−0.5 to −1.0)	Large positive (negative)
0.3 to 0.5 (−0.3 to −0.5)	Moderate positive (negative)
0.1 to 0.3 (−0.1 to −0.3)	Small positive (negative)
0.0 to 0.1 (0.0 to −0.1)	Negligible

Let's consider a few illustrative examples. A scatterplot of shoe size against the number of movies watched per year would likely look something like figure 6.2a. Knowing a person's shoe size gives no indication as to the number of

---

b. The exact nomenclature of the correlation strength and corresponding r-values vary somewhat in the literature. In this work, we will stick to the definitions given in table 6.1.

movies they watch. Why would it? A small positive correlation (r = 0.25) was found between an animal's life expectancy and its weight in one study.[3] This makes sense, since large animals, such as whales and elephants, can afford to take a longer time to mature and reproduce slowly since they are too big to be easily eaten by predators. Whereas smaller animals, such as a mouse, must mature fast and reproduce quickly before they are eaten. Such a small positive correlation would look something like figure 6.2b, where there is a general upward trend, but the correlation remains noisy and somewhat unpredictable. An example of a moderate positive correlation is that between muscle strength and height of men (r = 0.42), as illustrated in figure 6.2c.[4] An example of a large positive correlation is that between arm span and height of men (r = 0.99), as illustrated in figure 6.2d.[5] Clearly such a strong correlation is to be expected; the taller someone is the longer their arms tend to be. In fact, when the correlation is this strong one can fairly accurately predict the value of one variable simply by knowing the other.

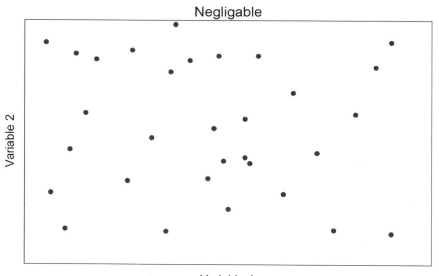

FIGURE 6.2a
Illustrative examples for each of the four positive correlation strengths: (a) negligible correlation, (b) small positive correlation, (c) moderate positive correlation, and (d) large positive correlation. Small, moderate, and large negative correlations would look like plots (b), (c), and (d), respectively, but would instead trend downward from top-left to bottom-right.

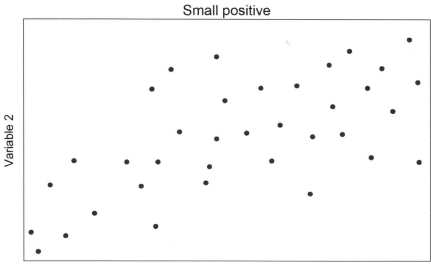

FIGURE 6.2b

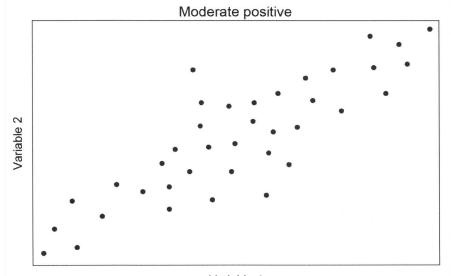

FIGURE 6.2c

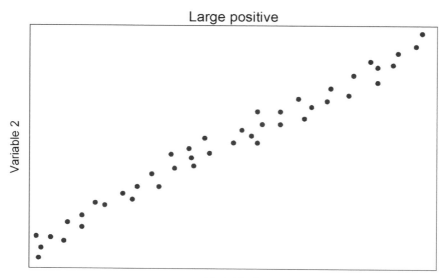

FIGURE 6.2d

Now we can return to our cigarette example with a better understanding. The correlation coefficient of r = 0.49 in this case, printed in the top left of figure 6.1, indicates a moderate (almost large) positive correlation. However, we only have a sample size of ten countries. What if these ten data points just happened to yield a correlation purely by chance? If our sample size is too small, we may be fooled into thinking there is a correlation when there isn't. Just how likely is it that the r-value determined from a small sample accurately reflects that of the entire population?

To answer this, we need something known as the p-value. A p-value is the probability that the correlation between two variables occurred by chance. The lower the p-value the better, since it lowers the chance of being misled by random statistical alignments. But how low must the p-value be before we accept that a particular correlation is real? Well, different criteria are used in different fields of research, but in this work, we define a threshold of $p \leq 0.01$, which means there is only a 1 percent probability (1-in-100) that results from a particular sample occurred by chance.[c] In our cigarette study, the p-value is found to be 0.15 as printed in the top left of figure 6.1, meaning there is a 15 percent chance that the stated correlation occurred by fluke. Somewhat unlikely, about a 1-in-7 chance, but you would not bet your house on these

---

c. Other common criteria include the 10 percent threshold, used in behavioral and social sciences, and the more restrictive 0.1 percent threshold sometimes used in the experimental sciences.

odds. Since this probability does not meet our standard of p ≤ 0.01 we must conclude that there is insufficient evidence that the correlation is real.

For Pearson's correlation coefficient between two variables to be reliable, several criteria should be satisfied, either exactly or in some cases approximately. These criteria are:

- *Data must be measured on an interval or ratio scale.*
  Did you know that twenty degrees Celsius is not twice as hot as ten degrees Celsius? This is because the Celsius temperature scale is an interval scale, not a ratio scale. When measuring temperature in Celsius, ratios are meaningless. This can be understood by thinking about the meaning of zero Celsius. Zero degrees Celsius is not the lowest possible temperature, you can easily make something colder than zero Celsius, just ask anybody in Siberia in January (minus twenty-five degrees Celsius on average). The temperature in Celsius can be negative. In other words, zero Celsius does not refer to a complete absence of temperature, it is an arbitrary definition based on the freezing point of water. However, it does make sense to say that twenty degrees Celsius is two degrees Celsius hotter than eighteen degrees Celsius. This is because it is meaningful to talk about temperature differences, or intervals, in Celsius. Hence, Celsius is an interval scale and not a ratio scale.

  However, there is another way of measuring temperature in which ratios do make sense; the Kelvin scale (K). For example, 20K *is* twice as hot as 10K. This is because zero in the Kelvin scale corresponds to a complete absence of the thing we are measuring, namely temperature. You cannot go below zero Kelvin. Nothing on Earth, or anywhere in the Universe, has a temperature below zero Kelvin (minus 273.15 degrees Celsius). Hence, Kelvin is a ratio scale.

  There are other scales, such as the nominal scale that distinguish items based on a qualitative description, such as an item's name. Taking an example from biology, the taxonomic rank—species, genus, family, and so on—defines a nominal scale. Pearson's correlation doesn't make much sense when we include such qualitative variables, since they don't have a numerical value associated with them. Therefore, for the correlation coefficient between two variables to be valid both should be defined on an interval or ratio measurement scale.
- *Measurement pairs.*
  This one is simple. Say we want to look at the association between the height of a person and their weight. For each individual we must make a pair of observations, both their height and weight need to be recorded. We cannot determine the correlation coefficient between a set of say one

hundred height measurements and thirty weight measurements. Each height value should have a corresponding weight value, corresponding to one pair of measurements for each individual in the study.
- *Measurement independence.*
Individuals observed in the study must define a completely independent measurement, not related to or dependent on any other individual. As an example, let's consider the association between height and weight again. If your sample includes several sets of siblings then the individuals observed do not constitute fully independent measurements, since siblings tend to be closer in height and weight to each other than to unrelated members of the study, due to genetic and environmental similarities. In this case, our data could be skewed by these dependent measurements, possibly leading to erroneous results.
- *Linearity.*
The two variables whose correlation we are interested in should have an approximately linear relationship. That is, the line of best fit should be straight. Pearson's correlation coefficient is not very reliable if the best fit to the data is some kind of curve.
- *No extreme outliers.*
Let's stick with our example of the correlation between height and weight. Measuring the height and weight of nineteen people might yield a large positive correlation coefficient of say r = 0.59. You begin to get excited, thinking you may have discovered a true correlation. However, the twentieth person may be so short and heavy that they single-handedly bring the r-value down to 0.28. This one outlier has significantly changed the result of the whole study, taking it from a large to a small positive correlation. Extreme outliers can skew results. When outliers are present in a data set, great care should be taken when interpreting the value and validity of Pearson's correlation coefficient. In extreme cases, or if erroneous measurements are suspected, it sometimes makes sense to remove outliers from the data set. However, one needs to think very carefully before simply omitting data that does not fit a preconceived hypothesis or the pattern followed by most data. As per the theme of this book, such anomalous results could be trying to tell us something important, even paradigm changing.
- *Homoscedasticity.*
This fancy word simply means that the spread of data points above and below the straight line of best fit must be approximately constant as you move along the line. This is more easily understood visually. Referring to figures 6.2c and 6.2d, we can see that figure 6.2d has a greater degree of homoscedasticity than figure 6.2c. This is because figure 6.2d main-

tains a constant thickness, or spread, as one moves along the trend line, whereas figure 6.2c does not. This property can be challenging to verify in practice, but it is often obvious from a scatter plot of the data.
- *Normality.*
A normal distribution, or informally a bell curve due to its resemblance to a bell, is a probability distribution that appears a lot in nature. Measuring the height of a large population of people typically yields a bell curve; some people are short, some are tall, but most stack up somewhere in the middle. This leads to the familiar bell shape, with a high central bump that smoothly tails off on both sides. The reliability of the Pearson correlation coefficient may also depend on one, or both, of the variables, being at least approximately normally distributed.[6] However, this is debated by the experts and there is no consensus on this requirement in the literature.[7]

In the real world, data is messy and is unlikely to exactly satisfy all these requirements. If your data is not perfectly linear, contains outliers, or does not follow a normal distribution then a different measure of correlation can be used that is not as sensitive to these requirements. This additional measure is known as Spearman's rank correlation coefficient. It is very similar to Pearson's coefficient but with one key difference. Let's look at a simple example to illustrate how Spearman's coefficient works.

Say the test scores for five high school students in physics and mathematics are:

Physics: 98, 10, 54, 34, 29
Math: 89, 19, 43, 47, 17

Instead of looking at the correlation between the exact scores in each subject, as we did above with Pearson's correlation coefficient, we might just want to know if the student who got the highest score in physics also got the highest in math. And if the student with the second highest score in physics got the second highest in math, and so on. In short, if you are good at physics do you also tend to be good at math? To do this we can rank the students in each subject, and then compare the respective ranks. For example, the first student in the list got the highest score in physics (98) and the highest in math (89). Therefore, they ranked 1st in both subjects—they're top of the class. The second student in the list ranked fifth in physics (10) but fourth in math (19). We can continue in this way, assigning a rank to each student in the two subjects. The result is summarized in table 6.2, where we have ordered the students according to their physics rank.

TABLE 6.2
A table of fictional test scores and their associated ranks. Data used to help illustrate Spearman's rank correlation coefficient.

Physics score	Physics rank	Math score	Math rank
98	1	89	1
54	2	47	2
34	3	43	3
29	4	17	5
10	5	19	4

A clear trend in rank can be seen from table 6.2; the better the student's rank in physics the better their math rank tends to be. The only slight deviation from this trend is in the last two rows of table 6.2, but even here the difference in rank is small. We can now make a scatterplot and work out the correlation coefficient, just as before. The only difference is that we are now looking at the correlation coefficient between the two rank lists. From this example, we can also see why Spearman's rank is not so sensitive to outliers. Consider the first student on the list, who got the highest score in physics and math. Their scores are roughly double everyone else's; they are something of an outlier. However, the ranking system does not care about whether they were fifty marks ahead of second-place or just one mark, all it records is the fact that this student was first-place. The information about how extreme an outlier they are is simply not encoded in the Spearman rank coefficient, making it a more reliable measure when outliers are present. The interpretation of the strength of Spearman's coefficient is considered identical to that of Pearson's coefficient, namely according to the definitions given in table 6.1.

In summary, we will explore several possible UFO correlations, searching for particularly strong trends and patterns that may help to crack this enigmatic phenomenon, or at least provide hints as to where to look in future studies. The results of each correlation study will be summarized by a scatter plot, including a trend line, Pearson's and Spearman's correlation coefficients, and their associated p-values. The reliability of each correlation study will be assessed with reference to the above criteria.

6.2 A UFO Database

Let's say you see something strange in the night sky and decide to dial 911 to report it. Or perhaps you contact NASA or the Federal Aviation Administration (FAA) or a local military facility. Emergency dispatch 911 centers throughout the United States and in many parts of Canada routinely redirect such reports to one place. As do NASA, the FAA, and military facilities. They end up at the National UFO Reporting Center (NUFORC).[8]

NUFORC was founded in 1974 in the United States and has since processed and catalogued more than ninety thousand reports, making it one of the largest UFO databases in the world. NUFORC is a nonprofit organization, which is important because it removes the financial incentive to sensationalize or even fabricate reported sightings. NUFORC's database is publicly available, and the reports are presented in the witnesses' own words. Some attempts have been made by NUFORC to remove obvious hoaxes and corroborate reports; however, it is impossible to do so in all cases given such a large database.

We begin by getting an overview of the data within the United States. The first question we may ask is, Which US states have the highest number of reported UFO sightings? To date, California has the most sightings with about 15,000, Florida is next with about 7,500, then Washington with approximately 6,700.[d] The distribution of UFO reports within the contiguous United States is best displayed visually, as can be seen in figure 6.3, where lighter shades of grey indicate more sightings and darker shades fewer sightings.

So, California has the most UFO reports, but it also has the largest population. Florida has the second largest number of sightings but also has a very large population (third biggest by state). Could it simply be the case that the more people there are the more UFO sightings reported? Well, we now have

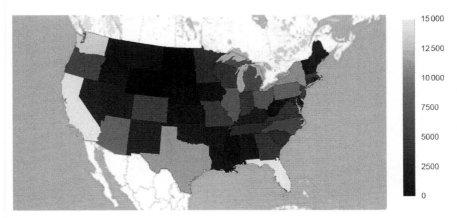

FIGURE 6.3
A heat map of the number of NUFORC reported UFO incidents for each state within the contiguous United States. The lighter the shade of grey the more UFO reports.

---

d. As of February 2022.

the tools to answer this question. Let's look at the number of UFO reports against the population for all fifty-one states (including Washington, DC). The results are shown in figure 6.4. The trend is clear: states with larger populations tend to file more UFO reports. The r- and p-values demonstrate that this is a statistically significant large correlation. Note that in figure 6.4, and hereafter, and will specifically denote the r- and p-values for the Pearson correlation test and and will refer to the r- and p-values for the Spearman test.

Cleary, we need to correct for population size. This is easy to do. We simply take the number of reports in each state and divide by that state's population, giving the number of reports per person (per capita). This correction enables a more accurate comparison between states and will hopefully allow any interesting features to become more apparent. The state with the highest number of reported UFO sightings per person is Washington, followed by Vermont and Montana. The distribution of UFO reports per capita within the contiguous United States is visually depicted in figure 6.5, where lighter shades of grey indicate more sightings per person and darker shades fewer sightings per person.

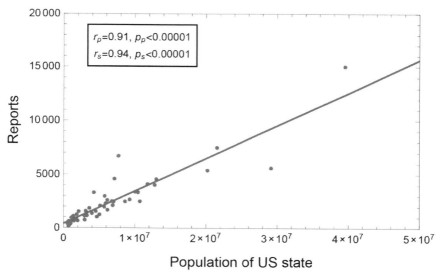

FIGURE 6.4
The correlation between the population of each US state and the number of UFO incidents reported to the National UFO Reporting Center (NUFORC). Both Pearson's and Spearman's correlation coefficients ($r_p = 0.91$ and $r_s = 0.94$) indicate a large correlation between population size and UFO reports. The associated p-values show this result is statistically significant at the 1 percent level.

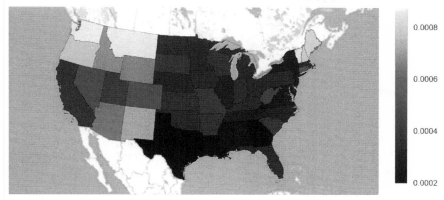

FIGURE 6.5
A heat map of the number of NUFORC reported UFO incidents per capita for each state within the contiguous United States. The lighter the shade of gray the more reports per capita.

Now we are left asking why? Why does the average person report more UFOs in the northwestern and northeastern states? Why are there proportionately fewer sightings per person in the central states, far from the coast? Could it be to do with geography or climate, race or wealth, pollution levels or military presence? There is only one way to find out; analyze more data.

6.3 Military

6.3.1 Introduction

At 10 p.m. on November 23, 1944, Lieutenant Edward Schlueter was piloting an allied plane behind German lines under the cover of darkness. The US operated plane was scouring the Rhine, searching in the darkness for Nazi convoys to bomb. The cockpit was dimmed for night vision. The pilot, radar operator and intelligence officer strained their eyes, anxiously looking for enemy silhouettes on the dark ground.

Out of the blackness, eight to ten bright orange lights appeared off the left wing, moving at tremendous speed. The onboard radar showed nothing. *Were enemy aircraft about to attack?* Ground control confirmed they saw no radar return either. Lieutenant Schlueter, a decorated veteran of nighttime combat, decided to turn and face the lights head-on. They instantly vanished. Then, moments later, reappeared in the far distance. This game continued for several minutes before eventually coming to an end.

The radar operator aboard the plane, Donald J. Meiers, was especially disturbed. During a debriefing after the incident, Meiers apparently slammed a comic book on the desk and yelled, "It was another one of those fuckin' foo fighters!" referencing a catchphrase from the comic strip. The name stuck. The flood of wartime sightings that followed would ever after refer to UFOs as *foo fighters*.[e]

Following World War II, almost every major conflict has been accompanied by many UFO sightings.[f] During the Korean war, at least thirty reports were made by US Air Force, Navy, and Marine Pilots. A significant proportion of these expert pilot observations were corroborated by airborne and ground radar.[9] The Vietnam War brought many UFO sightings, so much so that General George S. Brown, deputy commander for air operations, would receive daily briefings. Air Force intelligence officer Captain George Filer, who would provide these briefings, described a typical report in the following way:

> You'd have an aircraft flying along, doing around 500 knots [about 575 mph] and a UFO comes alongside and does some barrel rolls around the aircraft and then flies off at three times the speed of one of the fastest jets we have in the Air Force. So, obviously, it has a technology far in advance of anything we have.[10]

More recently, on the evening of March 19, 2003, unexplained lights were reported by civilians and paramilitary forces over Kirkuk, a city in the north of Iraq. The very next morning the Iraq War began when coalition forces invaded on mass. UFO sightings continued to be reported for the next two months.

A more specific connection between UFOs and the military has been proposed. After the mid-1940s, the UFO phenomenon skyrocketed. But why? Did the war-weary world fall into a kind of mass delusion involving strange lights in the sky? Or is there a reason UFOs began being reported in droves at this specific time in our history? Well, on December 2, 1942, the world's first man-made nuclear chain reaction occurred. This feat was a major milestone in the creation of the world's first atomic bomb, as well as a major milestone for mankind. The first detonation of a nuclear weapon occurred on July 16, 1945, during the code-named Trinity test in New Mexico. The atomic detonations over Hiroshima and Nagasaki followed only about a month later,

---

e. Incidentally, this is where the rock-band the Foo Fighters got their name.

f. UFO sightings have also been reported during large battles throughout history, although only after World War II were they properly documented. One such example dates to 74 BCE in what is now modern-day Turkey. The Roman army, led by Lucullus, was about to engage Mithridates of Pontus in battle when a huge flame-like object flew down between the two sides. The UFO was consistently described by both armies as being shaped like a wine vessel and molten silver in color.

on August 6 and 9, 1945, respectively. Is this timing merely a coincidence? Perhaps, but as we will see there have since been many incidents involving nuclear weapons and UFOs.

On March 24, 1967, Captain Robert Salas was sixty feet below ground in a nuclear launch bunker. It was the Cold War, and if required it was Captain Salas' job to launch any of the ten nuclear warheads Malmstrom Air Force base kept hidden below the Montana soil. Early that morning the phone rang. *Could this be the call he hoped would never come?* A guard posted above ground was on the other end of the line. The panicked guard told Salas that he could see a glowing red oval object that was about forty feet across darting around above the base, making ninety degree turns at high speed without a single sound. Relieved he would not have to launch the nukes; Salas woke his sleeping commander to tell him of the strange report. But relief quickly turned to panic. Suddenly, all ten nuclear missiles became inoperable. One-by-one, the lights on the control panel flicked off. Completely impotent, America's nuclear shield lay vulnerable for the next twenty-four hours during the height of the Cold War. A later investigation concluded that an externally generated signal had penetrated through the sixty feet of concrete and tampered with a specific component of the missiles. Following the incident, Captain Robert Salas was forced to sign a document confirming he would never speak about this encounter publicly, a promise he kept until it was declassified many years later.

Could it just be a coincidence that strange lights happened to be observed at the same time a fault shut down all the missiles? Well, it turns out that each of the ten nuclear warheads were operated and stored independently. They were designed that way so that if one went down the rest would remain operable. Yet, all ten simultaneously malfunctioned. The official government investigation into the matter concluded that the chances of all ten nuclear weapons simultaneously failing were "extremely remote." The unclassified report admitted that the investigation teams sent to Malmstrom were unable to determine a logical cause of the incident.[11]

Another nuclear-related incident took place on September 14, 1964. On this day, US Air Force First Lieutenant Robert Jacobs was assigned the task of capturing high-speed video of a test flight of the Atlas-D missile. The Atlas missile carried a dummy nuclear warhead. The main objective was to see if the warhead could be successfully launched into orbit from Vandenberg Air Force base in California. The video showed the missile climbing high over the Pacific Ocean until it reached a speed of eight thousand miles per hour. At this point in the video, Jacobs claims a disc-shaped craft entered the frame and fired four beams of light at the warhead while circling it. The warhead tumbled out of the sky. The next day, Major Florenze J. Mansmann ordered

Jacobs to his office. Inside were the major and two men dressed in grey suits. After reviewing the footage, the major ordered Jacobs to never speak of the incident and the men in grey suits took the footage and left. Jacobs never saw the footage again.

Many of the most credible and widely known UFO incidents also have a strong connection to the military. For example, the most famous of all, the Roswell incident, occurred about two hundred kilometers from the Trinity Site, where the first atomic bomb was detonated. The Lonnie Zamora incident took place just ninety kilometers from the Trinity site. The 2004 Nimitz incident is named after the lead vessel the USS *Nimitz*, a nuclear-powered supercarrier. These specific incidents suggest a possible correlation between the military and UFO sightings. But will this connection hold up on a national and international scale and under close scientific scrutiny? Let's analyze the data to find out.

6.3.2 Analysis

We begin by looking at general indicators of military presence and whether they exhibit any correlation with the number of reported UFO sightings per capita. Starting on a national level within the United States, we examine how the number of Air Force bases per square kilometer for each of the fifty-one states (including the District of Columbia) correlates with the number of reported UFO sightings per capita. We analyze the number of Air Force bases per square kilometer to yield a fairer comparison between smaller and larger states.

The results are shown in figure 6.6. We find the two variables to have a Pearson correlation coefficient of $r_p = -0.18$ and a p-value of $p_p = 0.20$. According to the criteria set out in table 6.1, this defines a small negative correlation. The relatively large p-value also shows this small correlation is not statistically significant at the 1 percent level. Checking these results against the Spearman coefficient, we find that in this case $r_s = -0.35$, which defines a moderate negative correlation. However, the p-value for the Spearman coefficient of $p_s = 0.013$ is also not significant at the 1 percent level. We therefore conclude that the military activity associated with the density of Air Force bases has no statistically significant effect on the number of reported UFO sightings per capita.

We now attempt to broaden our analysis in two ways. First, we widen our data collection to include countries around the globe. Second, we look at a broader indicator of military presence, namely the annual military budget of various countries. Our data comprises thirty countries in total, with ten of the countries having a large military budget, ten countries having a moderate budget, and ten countries having a relatively low military budget.[g] This sampling

---

g. See the appendix at the end of this chapter for a list of these thirty countries.

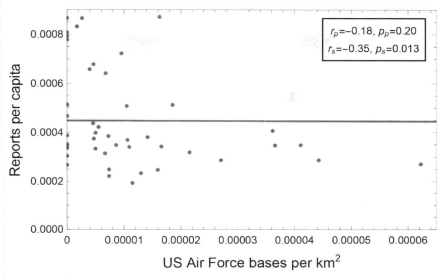

FIGURE 6.6
The correlation between the number of US Air Force bases per square kilometer and the number of reports per capita for all fifty-one US states (including Washington, DC). Pearson's r-value indicates a small negative correlation, while Spearman's r-value indicates a moderate negative correlation. However, the p-values in both cases confirm that there is no statistically significant correlation between these variables at the 1 percent significance level.

technique ensures a large range of data values which should enable a more accurate determination of any possible correlation. The results are displayed in figure 6.7. The Pearson correlation coefficient in this case is $r_p = -0.042$, indicating there is a negligible correlation between military spending and UFO reports per capita. The Spearman coefficient of $r_s = -0.43$ suggests a moderate negative correlation. However, for both measures of correlation the p-values show these results are not statistically significant at the 1 percent level.

Taking these two results together suggests that there is no sizeable correlation between the number of reported UFO sightings per capita and military presence or expenditure, at least in the specific cases studied. But perhaps UFOs are interested in a specific aspect of the military? As already indicated, a widely held speculation is that UFOs are for some reason especially interested in our nuclear weapons.

We begin by testing this idea within the United States. Nuclear test detonations have been conducted in six US states: Alaska, California, Colorado, Mississippi, Nevada, and New Mexico. By far the most nuclear tests have taken place in Nevada, with close to a thousand detonations. Figure 6.8a shows the number of nuclear tests per square kilometer against the number

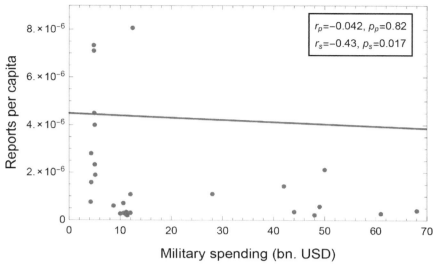

FIGURE 6.7
The correlation between the military budget in billions of USD and the number of reports per capita for thirty different countries. Pearson's r-value indicates a negligible correlation, while Spearman's r-value indicates a moderate negative correlation. However, the p-values in both cases confirm that there is no statistically significant correlation at the 1 percent significance level.

of UFO reports per capita for each of these states. Note that the data point for Nevada is off the chart due to the large number of nuclear tests (998), but it is included in the straight line of best fit and the calculation of the correlation coefficients and p-values. The Pearson correlation coefficient for the data shown in figure 6.8a is $r_p = -0.060$, with $p_p = 0.91$, demonstrating there is no statistically significant correlation as measured by the Pearson coefficient. However, when it comes to the Spearman rank correlation coefficient, there is a sizeable difference. Here, the result is $r_s = -0.54$, with $p_s = 0.27$. This measure suggests a large negative association; however, the large p-value shows it is not statistically significant. The sizeable difference between r_p and r_s is likely due to Nevada being an extreme outlier, coupled with the small sample size (since only six states have conducted nuclear test detonations). For further discussion of the reliability of this result, see appendix 6.8.2.

We now analyze UFO reports in countries with a nuclear arsenal. We investigate eight countries with nuclear warheads: Russia, UK, France, China, India, Pakistan, North Korea, and Israel. Figure 6.8b shows the correlation between the number of nuclear warheads a country has and the number of UFO reports per person. This data set yields $r_p = -0.16$ and $p_p = 0.70$, as well as $r_s = 0.048$ and $p_s = 0.91$, which verifies that there is no statistical correlation here either. Since figures 6.8a and 6.8b have limited sample sizes the results are likely to be unreliable. For more details on the reliability of the results presented in figures 6.8a and 6.8b, see appendix 6.8.2.

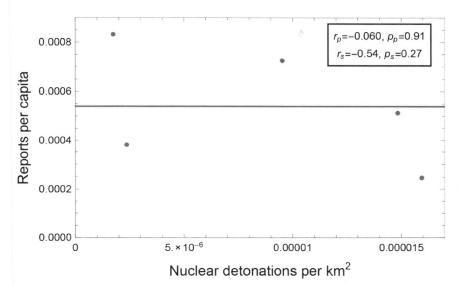

FIGURE 6.8a
(a) The number of nuclear test detonations per square kilometer against the number of reports per capita for the six US states. (b) The number of nuclear warheads against the number reports per capita for eight countries. The r and p-values indicate there is no statistically significant correlation for the data in graph (a) or (b).

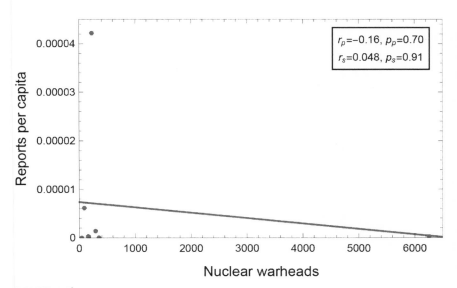

FIGURE 6.8b

In short, we find no statistically significant correlation between UFO sightings and any of the measures of military activity studied. It is important to stress, however, that this does not mean there is no connection in general. Other works report an association between UFOs and nuclear weapons, although as far as I can tell all of them are based on eyewitness accounts with little quantitative data.[12]

6.4 Water

6.4.1 Introduction

Water is miraculous. It is the only substance to exist as a solid, liquid, and gas on Earth. Another unusual property of water is that its solid form, ice, is less dense than its liquid form, water. This is important because it allows ice to float on water. If it didn't, there could be no white polar ice caps to reflect the sun's light—and we would all bake to death. Also, many of the world's oceans would freeze solid, from the bottom up. Water is sometimes called the *universal solvent* because so many different substances dissolve in it. This feature is vital to life on Earth as it helps transport important substances like oxygen and nutrients around the body. In short, water is a very unusual and important compound, and on Earth it is present in all three states. Perhaps UFOs travel here to study our beautiful blue planet?

Another proposal stems from the ubiquity of water on Earth. Of Earth's surface, 71 percent is covered in water. Current estimates put the average ocean depth at about 3.7 kilometers, with the deepest parts descending nearly eleven kilometers. Only about 10 percent of the ocean floor has currently been mapped with high resolution. What's down there, in the other 90 percent, is a complete mystery. The vast extent and depth of the oceans, coupled with our ignorance of it, has led many to suggest it is the perfect hiding place for extraterrestrial visitors that do not want to be seen by the indigenous population.

Alternatively, instead of hiding under water, the answer could be hiding within water itself. As everyone knows, water has the chemical formula H_2O, meaning it is made of two hydrogen and one oxygen atoms. We can separate hydrogen from oxygen via a process known as the electrolysis of water. Hydrogen is an excellent fuel source. One example application is the hydrogen fuel cell used to generate clean electricity. Could UFOs be harvesting our water for energy? It is unlikely that any extraterrestrial civilization would visit Earth with the sole purpose of seeking water as a fuel source, since ice is common throughout the cosmos, but it is conceivable that they would use it as a kind of galactic refueling station on their way across the cosmos. In fact, NASA have explored similar refueling strategies themselves.[13]

Where Are All the UFOs?

Possible motivations aside, the idea that UFOs are somehow connected with water primarily originates from the large number of sightings and encounters that have taken place either on, near, or even under, a large body of water. In fact, there have been so many that researchers coined a new term, unidentified submerged object (USO), to distinguish them from familiar UFOs. For instance, a total of 623 UFO/USO sightings over the Atlantic and Pacific oceans were reported by the US Navy to the US governments Project Blue Book alone. Many of these sightings were officially declared unidentified by the US government.

At around midnight on October 4, 1967, Shag Harbour, a small sleepy fishing village in Nova Scotia, Canada, became the epicenter of one of the best documented UFO/USO cases in history. The remarkable series of events that would unfold that night began about four hours prior to this, at 7:15 p.m. local time. At this time, Air Canada Flight 305 was at an altitude of twelve thousand feet flying approximately thirty miles east of Montreal, en route to Toronto. Visibility was mostly clear with thin wispy clouds below the plane. First Officer Robert Ralph and Captain Pierre Charbonneau noticed a bright object to the south (their left). The object was rectangular, orange in color, and had a string of smaller lights trailing it. The strange object was maintaining a distance of a few miles on a course parallel to their Douglas DC8 plane and about twenty degrees above the horizon. At approximately 7:19 p.m. they witnessed a large but strangely silent explosion next to the object. The explosion turned into a spherically shaped white cloud, before turning red, then violet, and then finally blue. At 7:21 p.m., a second, much larger, explosion occurred that also eventually turned into a blue cloud. These specific colors will become important later. Upon landing, the first officer and captain submitted official reports on the incident, despite fearing for their reputations and careers.

Later, at 9 p.m., Captain Leo Howard Mersey and his crew of nearly twenty men aboard the sea vessel MV *Nickerson* witnessed four bright objects in a rectangular formation in the northeastern sky. Their boat was floating thirty-two nautical miles south of Sambro, Nova Scotia. Mersey confirmed four sharp returns on his ships radar about sixteen miles to the northeast. Just to keep tabs, this case now includes multiple trained and untrained visual witnesses as well as corroborating radar returns. Captain Mersey radioed the rescue coordination center (RCC) and the harbor master in Halifax asking if there were any navy training exercises in the area. There were not. Upon returning to port, Mersey filed a report with the Lunenburg Royal Canadian Mounted Police (RCMP), which is publicly available on the National Archive of Canada (file RCMP 67-400-23-X). Note that near the bottom of this official document the RCMP state that:[14] "Captain Mersey is considered to be a reliable type individual and bears a good reputation in his community."

But the encounter does not end there. Shortly before 11:20 p.m., at least eleven people witnessed a brightly lit object descend toward Shag Harbour. The object, whatever it was apparently in trouble. Witnesses report a high pitch whistling sound, "like a bomb" descending. Then a loud bang. Local resident Lorie Wickens and four others saw an object floating approximately three hundred miles offshore. Wickens, thinking an airliner had crashed, immediately phoned the RCMP based in Barrington Passage. Two RCMP officers, Ron O'Brien and Ron Pond, arrived at the scene within fifteen minutes. The object they saw was now about half a mile offshore and was still emitting light. The police officers contacted the RCC in Halifax to ask if they had any missing aircraft. They then stood and watched in silent resignation as the object slowly sank below the waves, vanishing from view.

Concerned there were survivors of the presumed plane crash, locals took to their fishing boats and went out to try to rescue them. Strangely, they found absolutely no wreckage. An airliner, or even a light aircraft or helicopter, that crashes into the sea usually leaves substantial debris. Yet, no debris was found. However, what they did find was a thick blanket of yellow foam covering the sea, about eighty feet wide and half a mile long. It smelled strongly of sulfur— like rotten-eggs. Interestingly, sulfur also burns with a bright blue flame, possibly explaining the blue cloud associated with the explosions reported earlier by the crew of Air Canada Flight 305.

Approximately an hour later, a Canadian Coast Guard cutter arrived at the scene. It also found zero wreckage. Two days later, navy divers from Fleet Diving Unit Atlantic arrived and began searching the seafloor for the next three days. They found nothing. The RCC at Halifax later confirmed that no private, commercial, or military aircraft were missing anywhere along the entire eastern seaboard. Whatever crashed in the waters off Shag Harbour that night was likely not a terrestrial aircraft.

The Shag Harbour incident is just one of many involving water. But rather than discussing cases on an individual basis, it may be beneficial to analyze the possible connection between UFOs and water on a larger scale. Is there a real, statistically significant, connection with water? If so, does it hold on national and global scales? What type of water? Does it make a difference if it is freshwater or seawater? Let's find out.

6.4.2 Analysis

We begin on a national level, looking at the possible connection between water and UFO reports within the United States. Going back to our heat map of the contiguous United States (figure 6.5) we can see a pattern. The states with the highest number of sightings per capita tend to be on the east and west coast, with few reports in the middle of the country. If there really are more UFOs observed on each coast, then we should see the number of reports increase as we move away from the geographical center of the US (located in Kansas) along lines of constant latitude (i.e., east-west distance). To some extend this is in fact what we see, as shown in figure 6.9. Pearson's coefficient for this data indicates a moderately strong positive correlation, but the p-value shows it is not a statistically significant correlation at the 1 percent significance level. Yet, the p-value of 0.011 could not be any closer without being statistically significant. The result of the Spearman correlation test also corroborates a moderate positive correlation with a low p-value, although again not quite low enough to be declared statistically significant according to our criteria. This result is close enough to significance to warrant further study.[h]

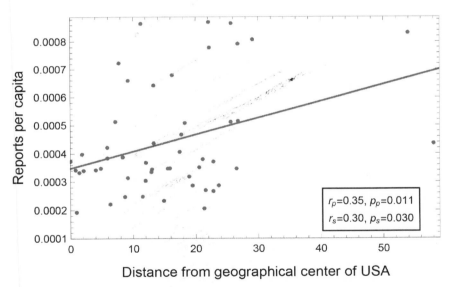

FIGURE 6.9
The distance along lines of constant latitude from the geographical center of the contiguous USA (near Lebanon, Kansas) against the number of UFO reports per capita for all fifty-one states of the USA. The Pearson and Spearman r-values suggest a moderate positive correlation; however, the p-values demonstrate they are not quite statistically significant.

---

h. One possible approach could be to analyze the physical distance to the coast for a sample of individual sightings, rather than only the east-west distance of each state.

Following this enticing result, let's look in more detail at the impact the coast has on UFO reports. In figure 6.10 we plot the coastline length for each US state that has a coastline (thirty out of fifty-one states) against the number of UFO reports per capita. Pearson's coefficient suggests a moderate positive correlation, but Spearman's coefficient does not agree and instead indicates a negligible correlation. Moreover, neither p-value demonstrates statistical significance at the 1 percent level. For a discussion of the reliability of these results see appendix 6.8.3.

We now broaden our investigation to countries around the world. In figure 6.11, we plot the coastline length against the number of UFO reports per capita for twenty-nine countries spread across the globe. Here there is a clear upward trend, the longer a countries coastline, the more UFO sightings there are. The numbers support this association. Pearson's r-value suggests a large positive correlation, and the p-value for this statistical test indicates the correlation is statistically significant at the 1 percent level. In fact, the very low p-value of $p_p = 0.00027$ indicates there is only about a 1-in-3,700 chance of the

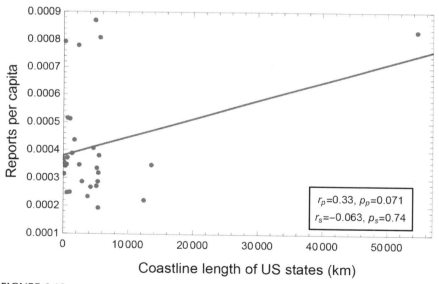

FIGURE 6.10
The length of coastline of all US states with a coastline (a total of 30 states) against the number of UFO reports per capita. The r-values suggest either a moderate positive correlation or a negligible correlation, but the p-values demonstrate the correlation is not statistically significant at the 1 percent level.

correlation not being real. A compelling result. The results for the Spearman correlation test are not quite as good. In this case, a moderately strong positive correlation is found and the p-value of $p_s = 0.020$ just misses out on statistical significance. Again, this result is so tantalizingly close to significance it demands further investigation.

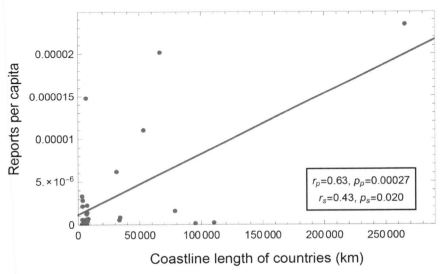

FIGURE 6.11
The length of coastline in kilometers against the number of UFO reports per capita for twenty-nine countries. Pearson's coefficient implies a large positive correlation, and the p-value indicates a statistically significant result. However, the Spearman coefficient suggests only a moderate correlation that is not quite statistically significant at the 1 percent level.

Can we go further? Let's try to discern whether it's saltwater or freshwater that has the greatest impact. To this end, we again begin by looking at just the United States. In figure 6.12 we plot the fraction of the total area that is covered by freshwater against the number of UFO reports per capita for all fifty-one US states. There is no significant correlation here. Making the same analysis but for thirty countries around the world also shows no significant correlation, as can be seen in figure 6.13. So, there seems to be a stronger connection between UFOs and coastal seawater, than between UFOs and freshwater. Why this is the case remains a mystery.

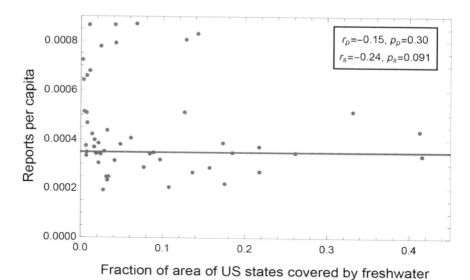

FIGURE 6.12
The fraction of the total area that is covered by freshwater against the number of UFO reports per capita for all fifty-one US states. The r and p-values show there is no statistically significant correlation.

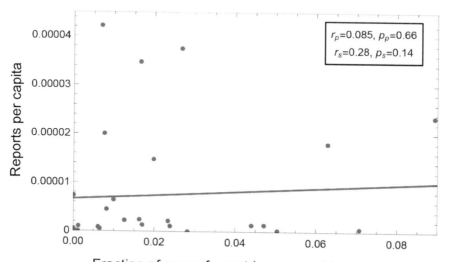

FIGURE 6.13
The fraction of the total area that is covered by freshwater against the number of UFO reports per capita for 30 different countries. The r and p-values show there is no statistically significant correlation.

6.5 Environment

6.5.1 Introduction

On September 16, 1994, at 10 a.m. the children of Ariel school in Zimbabwe were outside playing at recess. The children spotted several silver discs glinting in the morning sky. As witnessed by no less than sixty-two children aged between six and twelve years old, the objects descended to a field just beyond the schoolyard. Upon landing, beings got out of the disc and walked towards the children. The humanoid-looking beings are consistently described by the children as being small men dressed in black overalls with large eyes shaped like rugby balls. The entities, whatever they were, then proceeded to telepathically convey environmental warnings to the children.

John Edward Mack, a Harvard University professor, and head of the Psychiatry Department at Harvard Medical School, as well as the winner of the 1977 Pulitzer Prize, was intrigued by this case and decided to interview the sixty-four witnesses. One child told Mack they received a message from the humanoid beings that "pollution mustn't be", while another eleven-year-old recounted that "I think they want people to know that we're actually making harm on this world." A third witness explained the message she received in the following way during an interview with Professor Mack:[15]

> *Girl*: "What I thought was that maybe the world's gonna end, they're telling us the world's gonna end. Maybe because we never . . . don't look after the planet and the area properly. I just felt all horrible inside."
>
> *Mack*: "Say more about that horrible feeling, what was it like?"
>
> *Girl*: "It was like in the world all the trees would just go down and there will be no air and people will be dying."

This story made headlines around the world. Many have dismissed the 1994 incident as an example of mass hysteria. However, in an investigation by Dr. Mack, the children were found to not have much prior knowledge of UFOs or popular UFO perceptions and were remarkably consistent in almost every detail. Judy Bates, a teacher at the time of the alleged encounter and now a headmistress, struggled for years with what occurred that day. Finally, during an interview for the 2020 movie *The Phenomenon*, Judy opened up about her experience. Pressed to summarize what happened that September morning in just one sentence she simply said, "Alien's visited us."[16]

So, are extraterrestrials visiting Earth to warn us of the damage we are inflicting on our home planet via CO_2 emissions, nuclear waste, and pollution in general? Due to this incident and other similar experiences, this idea has become a popular theme in UFO research over the decades, with subsequent research indicating some interesting possible connections. For example, one particularly rigorous study by a research group at the Toulouse School of Economics looked at unidentified aerial phenomena observed in France over a sixty-year period using sophisticated spatial point pattern techniques.[17] The group discovered that sightings of unidentified aerial phenomena (UAP) (essentially a different name for UFOs) correlated with a few environmental factors. For example, they managed to establish a statistically significant link (p-value = 0.00013) between nuclear activities and unidentified aerial phenomena in France, something that has been posited for decades. A strong relationship between UAPs and contaminated French land (p-value = 0.00542) was also revealed for the first time in this study.

Since this study was specific to France, it would be interesting to broaden its scope by including correlation studies focused on other individual countries as well as a meta-study including countries from all over the globe. In the following section, we attempt to do just that.

6.5.2 Analysis

The United States of America was the world's first atomic nation. As the world's richest country, and only superpower, it also has a huge energy demand. With less than 5 percent of the world's population, the US consumes almost 16 percent of the world's energy. Some of this substantial demand is met by commercial nuclear reactors that use uranium oxide to generate nuclear power through fission reactions. Inevitably, the nuclear fuel used in these reactors degrades over time, eventually becoming unable to produce significant amounts of power. This so-called spent nuclear fuel (SNF) must be removed and carefully stored at various sites around the United States. The United States of America also has the largest number of UFO reports of any country in the world, by a significant margin.

For these reasons, it makes sense to first analyze the possible association between nuclear storage sites and UFO reports within the United States.[18] To correct for population size and land area we plot the number of nuclear storage sites per square kilometer against the number of UFO reports per capita. The results of this study are presented in figure 6.14. Even by visual inspection, it is obvious that there is no clear trend. The data points are scattered evenly over the plot area. The trend line supports this, with an almost perfectly horizontal best fit line. More quantitatively, Pearson's r-value shows a negligible correlation ($r_p = -0.097$) and the associated p-value ($p_p = 0.50$) confirms it is not statistically significant. Spearman's r-value suggests a small negative correlation ($r_s = -0.28$) but one that is also not statistically significant ($p_s = 0.048$) according to our standard.

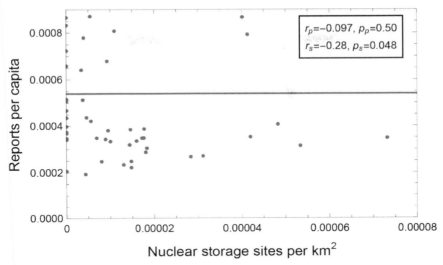

FIGURE 6.14
The number of nuclear storage sites per square kilometer against the number of UFO reports per capita for all 51 states of the USA. The r-values show a negligible or small negative correlation, and the p-values confirm there is no statistical significance in this case.

So, there is no clear correlation within the United States for these specific variables. But perhaps a pattern might emerge if we widen our analysis to the international level. The International Atomic Energy Agency (IAEA) estimates that between 11,000 and 18,000 tons of spent nuclear fuel are produced worldwide every year. As of 2013, this amounts to approximately 370,000 tons of SNF worldwide since the first reactor was connected to the grid.[19] Today, this staggering value is higher still. The damage these huge quantities of nuclear waste must be inflicting on our environment is concerning, to say the least.

We now analyze the possible association between the amount of spent nuclear fuel and the number of UFO sightings on an international scale. Due to restrictions on the availability and reliability of nuclear waste data in certain parts of the world, we are unfortunately restricted to a data set of just sixteen countries all located within Europe.[20] Nevertheless, this sample should be sufficiently diverse and large enough to at least capture the most salient features. Figure 6.15 shows a plot of spent nuclear fuel (in tons) against the number of UFO reports per capita for sixteen European countries. A list of the sixteen countries analyzed can be found in appendix 6.8.4. Running the Pearson correlation test gives a test statistic of $r_p = 0.18$, indicating a small positive correlation. However, the p-value for the Pearson test yields $p_p = 0.51$, which proves it is far from being statistically significant. The Spearman test yields r_s = −0.27 and p_s = 0.31, which confirms the lack of statistical significance. The noticeable difference between the Pearson and Spearman test results may be due to outliers in the data, for example the anomalous point far above the trend line in figure 6.15 (see appendix 6.8.4 for more details).

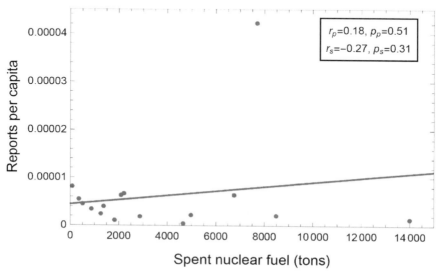

FIGURE 6.15
The amount of spent nuclear fuel (in tons) against the number of UFO reports per capita for sixteen European countries. The r and p-values indicate there is no statistically significant correlation between these variables.

Carbon dioxide (CO_2) emissions are widely regarded as the main contributor to global climate change. China, the United States, and India emit the most CO_2 per year. In fact, these three countries combined account for a shocking 40 percent of the world's total emissions. In figure 6.16 we plot CO_2 emissions (as a percentage of the world total) against the number of UFO reports per capita for thirty different countries distributed around the world. For a full list of the thirty countries studied, see appendix 6.8.4. CO_2 emissions for China and India are too large to be reasonably displayed but are included in calculations of the r-values and associated p-values. Since we already studied nuclear waste on a national level within the United States, we do not include the United States in this study. Pearson's r-value for this data indicates a small negative correlation (r_p = −0.13), however the p-value (p_p = 0.48) shows that it is not a statistically significant correlation. Spearman's rank coefficient (r_s = −0.042) suggests an even weaker negligible correlation, with the p-value again confirming no significance.

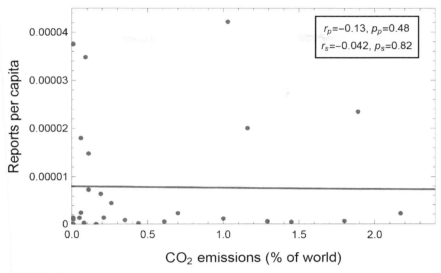

FIGURE 6.16
CO2 emissions as a percentage of the world total against the number of UFO reports per capita for thirty countries. The r-values suggest a negligible or small negative correlation, but the p-values show no statistical significance.

At least based on these results, there does not appear to be any sizeable association between measures of environmental damage and UFO reports.

6.6 Earthquakes

6.6.1 Introduction

Earthquakes and UFOs have a long history. In the year 869 AD, a massive earthquake, with an estimated magnitude of at least 8.4 on the moment magnitude scale, devastated ancient Japan. An official account of the disaster was recorded in the historical text *Nihon Sandai Jitsuroku* and reads:

> On the 26th day of the 5th month (9 July 869 AD) a large earthquake occurred in Mutsu province with some strange light in the sky. People shouted and cried, lay down and could not stand up. Some were killed by the collapsed houses, others by the landslides. Horses and cattle got surprised, madly rushed around and injured the others. Enormous buildings, warehouses, gates, and walls were destroyed. Then the sea began roaring like a big thunderstorm. The sea surface suddenly rose up and the huge waves attacked the land. They raged like nightmares, and immediately reached the city center. The waves spread thousands of

yards from the beach, and we could not see how large the devastated area was. The fields and roads completely sank into the sea. About one thousand people drowned in the waves, because they failed to escape either offshore or uphill from the waves.[21]

As tragic as this account is, it is exactly what one would expect from such a devastating earthquake, except for one thing. Look again at the first sentence: "some strange light in the sky"? This is an unusual remark. If we only had this one example, it would be easy to dismiss it as a historical transcription error or mistaken observation. However, there have been myriad reports of strange lights associated with earthquakes over the centuries. For instance, before the 1811/1812 New Madrid earthquakes in Missouri, numerous residents reported seeing bizarre lights in the air, and before the great San Francisco earthquake of 1906, scores of people reported seeing a faint rainbow of light.

But there's more to support this connection than just anecdotal records from history. There is a body of published literature studying the appearance of luminous phenomena around earthquakes, although mainly within localized regions of the United States. For example, one large study including more than twenty thousand UFO reports more than fourteen years found that the number of seismic events within the northeastern, eastern, and central regions of the United States had a strong temporal correlation with UFO report numbers.[22] However, correlations between seismic numbers over the whole of the United States and UFO reports were not statistically significant.[23] In yet another study, a strong temporal correlation was found between UFO report numbers and seismic activity specifically within the Uinta Basin in Utah.[24]

Similar observations continue to the present day. On the night of September 7, 2021, a magnitude 7.1 earthquake shook Acapulco, a town on Mexico's Pacific coast. During the quake, many people reported seeing bizarre bright flashes of light in the dark Mexican sky. Many residents took to social media, declaring "the apocalypse is upon us." After the panic had subsided, some of these flashes were shown to have simple prosaic explanations, such as exploding electrical transformers. However, the mysterious luminous phenomenon that has been reported alongside earthquakes throughout history was also thought to be partly responsible. Modern scientific research into this phenomenon is ongoing, and much remains unknown, but at least the phenomenon now has a name: *earthquake lights*.

Earthquake lights, or EQLs for short, are defined as any earthquake-related luminous phenomena. These elusive lights have not only been reported during seismic activity but also in the hours and days before an earthquake. For example, for nearly two weeks before a magnitude 5.8 earthquake in Quebec, Canada, people reported a bright purple-pink ball of light over the nearby St. Lawrence River. And as long as nine months before a 2009 magnitude-6.3

earthquake struck L'Aquila, Italy, flame-like lights were observed flickering above the streets and valleys of the region. Due to this feature, it may even be possible to use EQLs as a kind of early warning system, alerting authorities to an imminent seismic event and potentially saving lives.

Although there is now a consensus among scientists that the phenomenon of earthquake lights is real, its cause is still largely a mystery. The most likely explanation is electromagnetic discharge due to friction as tectonic plates grind against one another during an earthquake. However, the fact that earthquake lights are so rare and unpredictable, and appear in such a diverse range of colors, shapes, and sizes, has led many to speculate that they could be connected to UFOs. Perhaps extraterrestrials are concerned for our safety and are monitoring the earthquakes. Or perhaps they have a keen interest in seismic activity and wish to observe it up close and in detail.

On the other hand, if EQLs are natural phenomena, they could offer a rational explanation for UFO sightings, without the need for anything supernatural or extraterrestrial. Referring to our heat map of UFO sightings in the United States (figure 6.5), notice that the states with the highest number of UFO sightings per person are concentrated in the west and northwest regions, exactly where most seismic activity occurs in the contiguous United States. The collision between certain types of tectonic plate boundaries necessarily occurs more often at the edge of continents. Thus, specific types of plate collisions should occur more frequently along continental coastline. It is then possible that this effect may contribute to the correlation between UFO sightings and coastline length partially established previously.

It is also entirely possible that EQLs are a hallucination. Earthquakes are known to produce very strong electromagnetic fields, and strong electromagnetic fields are known to cause hallucinations. An example of this is known as transcranial magnetic stimulation (TMS). Since it was invented in the 1980s, neuroscientists have used TMS to investigate how the brain functions. The procedure involves focusing powerful and rapidly changing electromagnetic fields on specific parts of the brain. When this field is applied to the visual cortex subjects report seeing various luminous phenomena such as disks, lines, and balls of light. Move the field slightly within the patient's cortex and the lights appear to move. Also, another study demonstrated that minor earthquakes contribute to psychotic disorders, hypothesizing that this is due to the strong electric fields present during earthquakes or associated ultra-low-frequency sound emissions.[25]

In any case, we can examine the possible correlation between earthquakes and the number of UFO reports on a national and international scale to try to tease out any such connection.

6.6.2 Analysis

We begin our analysis within the United States. We will examine the number of earthquakes of magnitude 3.5 or greater, to encapsulate any possible effects due to minor earthquakes. This seismic data is studied for all fifty-one states. Since states vary widely both in size and population, we correct for this by plotting the number of earthquakes per square kilometer against the number of UFO reports per capita. The results of this analysis are shown in figure 6.17. Pearson's correlation coefficient for these two variables is $r_p = 0.050$, making it a negligible correlation. The p-value for Pearson's test is $p_p = 0.73$, which is far from defining a significant association. On the other hand, the results are very different for Spearman's rank coefficient. For this measure, we obtain $r_s = 0.48$, which defines a moderate (very nearly large) positive correlation. The p-value for the Spearman test of $p_s = 0.00032$ establishes this measure as statistically significant at the 1 percent level. There are several reasons to think the Spearman test is the more reliable one in this case, as discussed in appendix 6.8.5. Since both measures of correlation do not yield statistical significance, we must refrain from making any substantial claims here. Nevertheless, this is an intriguing result that warrants further investigation.

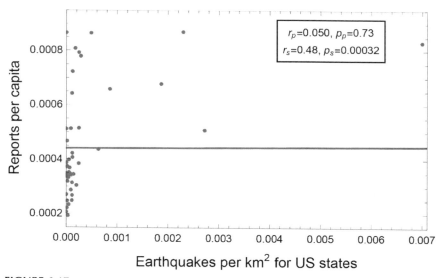

FIGURE 6.17
The number of earthquakes (magnitude 3.5 and greater) per square kilometer against the number of UFO reports per capita for all fifty-one US states. Pearson's test yields a negligible statistically insignificant correlation. However, Spearman's test yields a statistically significant moderate positive correlation.

To pursue this suggestive result further we look at the same correlation but this time on a global scale. More specifically, we analyze the number of strong earthquakes, defined to have a magnitude of 7.0 or greater, per square kilometers against the number of UFO reports per capita for eighteen different countries from around the world. Here we analyze stronger earthquakes since data on minor earthquakes is not readily available for some countries (for a list of the eighteen countries analyzed see appendix 6.8.5). The results of this analysis are shown in figure 6.18. This broader study yields a statistically insignificant negligible correlation according to both a Pearson and Spearman test. Therefore, the previously obtained partial correlation within the USA now seems less likely to be significant considering the results from this broader study. Or it could be that smaller magnitude earthquakes have a greater association with UFO sightings than stronger earthquakes. It would be interesting to reproduce figure 6.18 for smaller magnitude earthquakes to explore these ideas in future studies.

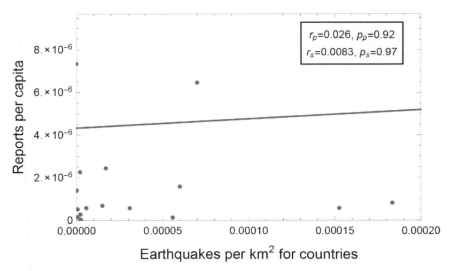

FIGURE 6.18
The number of strong earthquakes (magnitude 7.0 and greater) per square kilometer against the number of UFO reports per capita for eighteen different countries. Both sets of r and p-values indicate a statistically insignificant negligible correlation.

Of course, the results presented here do not conclusively show there is no connection between UFOs and earthquakes. They only demonstrate there is no clear statistical correlation for the specific parameters studied. This is especially important to point out, given the existing body of literature that has established statistical connections between UFO sightings and earthquakes within their study-specific contexts.

6.7 Blood

6.7.1 Introduction

The death of more than 100,000 fetuses and newborn babies each year may be linked to an unusual feature of human blood.[26] The disease thought to be responsible for these tragic deaths is called Hemolytic Disease of the Newborn (HDN) and is a result of the mother and baby having incompatible blood groups.

The four main blood groups are A, B, AB, and O. Each group can be either rhesus positive (Rh-positive) or rhesus negative (Rh-negative). The rhesus factor refers to a type of protein found on the covering of red blood cells. If you have this specific protein, you are Rh-positive. If you do not, you are Rh-negative. A mother with Rh-negative blood carrying a fetus with Rh-positive blood can cause significant and life-threatening problems during pregnancy. In this case, the mother's immune system sees the baby's blood as foreign. Antibodies are produced in the mother that sadly attack and often kill the unborn baby. The mother's body does not recognize the embryo as human, it is seen as a foreign body to be expelled, just as if the baby was from a different species than the mother.

From an evolutionary point of view, one would expect that such a genetic trait would have long ago been removed from the gene pool via natural selection. Yet, about 15 percent of the world's population has Rh-negative blood. Why? Does Rh-negative blood convey some genetic advantage that compensates for the increased number of fetal deaths? A possible clue comes from sickle cell anemia (SCA), a genetic mutation that affects red blood cells predominately in people of African descent. Since SCA prevents most of its victims from living beyond fifty years old, it was also puzzling why it was so prevalent in the gene pool. The answer came in 1954 when it was shown that SCA had a protective effect against malaria. The clear advantage bought by this mutation outweighed its disadvantage. It's currently unknown whether a similar advantageous effect may result from having Rh-negative blood, although some studies indicate subjects with Rh-negative blood have reduced ill-effects from certain viruses and parasites.[27]

Deletion of the gene responsible for Rh-negative blood, referred to as RHD deletion, occurred sometime during the early evolution of hominids in Africa,[28] and almost certainly before anyone migrated out of Africa.[29] Similar genetic mutations are relatively commonplace in our history. However, given such a clear genetic disadvantage, how the Rh-negative type became so established in a predominantly Rh-positive population is a complete mystery to modern science.[30]

The Basque population located in the Pyrenees Mountains between Spain and France are approximately 35 percent Rh-negative, among the highest percentages in the world. The language spoken by the Basque people is also unusual. It is thought to be a language isolate, meaning that it has no known connection to any other language. It is also a very old language, predating the spread of Indo-European languages throughout Europe. Since the colonial period, Basques have spread across the globe, with significant populations in the United States, Canada, Chile, Mexico, and Australia.

What does all this have to do with UFOs? Well, the numerous mysteries surrounding the Rh-negative blood group have sparked wild speculation that extraterrestrials may have tampered with our genetic code during our early evolution. The idea is that the 15 percent of the population having Rh-negative blood are genetically distinct from normal human beings, a hybrid species if you will. This is one of the more crazy-sounding claims, but due to its popularity in the UFO community, it will be investigated. Surely this idea can quickly be squashed, and we can move on to more sensible studies, right? Well, not so fast.

6.7.2 Analysis

As mentioned in section 6.7.1, approximately 15 percent of the world's population has Rh-negative blood. However, this percentage varies significantly throughout the world. For example, less than 1 percent of the Chinese population is Rh-negative, whereas approximately 29 percent of Moroccans living in the high Atlas Mountains are Rh-negative. We may exploit this variation by testing whether countries with proportionately larger Rh-negative populations report more UFO sightings.

We conduct this study using a sample size of fifty-one countries.[31] To conduct a broad and reliable investigation we sample countries over the entire range of Rh-negative percentages. Figure 6.19 shows the number of reports per capita against the percentage of the population having Rh-negative blood for fifty-one countries from around the world. Remarkably, figure 6.19 shows a large positive correlation in both the Pearson and Spearman tests ($r_p = 0.50$ and $r_s = 0.60$). Furthermore, both test statistics have p-values ($p_p = 0.00016$ and $p_s < 0.00001$) that show statistical significance well below the 1 percent level.[i] A striking result. Various cross-checks and tests of these results can be found in appendix 6.8.6.

What can we conclude from this? Does it mean extraterrestrials in their UFO craft are more frequently visiting their genetically engineered human

---

i. The probability that the Pearson correlation is due to chance alone is about 1-in-6,250.

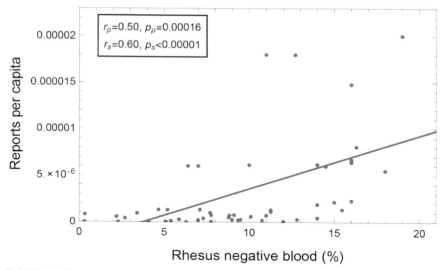

FIGURE 6.19
The population percentage with rhesus negative blood against the number of UFO reports per capita for fifty-one different countries. Both r-values suggests large positive correlations and their respective p-values establish statistical significance at the 1 percent significance level.

hybrids? Probably not. When interpreting correlation studies, such as this one, great care must be taken. Just because two variables are correlated does not mean that one causes the other. This fallacy is enshrined in the expression that *correlation is not causation*. Nevertheless, this result is interesting and understanding its origin and meaning is worth more careful study.

In such a future study, it may be worth examining whether patients with Rh-negative blood may be more likely to hallucinate such UFO sightings than the average person since Rh-negative blood has been broadly linked with mental health issues in the existing literature.[32] It would also be interesting to investigate whether some of the UFO sightings might be culturally dependent. For example, if the Basque culture was steeped in UFO mythology the propensity for seeing UFOs may have spread with the Basques throughout the world, along with their Rh-negative blood.

6.8 Appendix

6.8.1 Statistical Calculations

Pearson's correlation coefficient r_p for a sample of size n is defined by

$$r_p = \frac{\sum_{i=1}^{n}(x_i - \bar{x})(y_i - \bar{y})}{\sqrt{\sum_{i=1}^{n}(x_i - \bar{x})^2}\sqrt{\sum_{i=1}^{n}(y_i - \bar{y})^2}}, \qquad (6.1)$$

where (x_i, y_i) are the coordinates of each data point in the sample and $\bar{x}$ and $\bar{y}$ are the mean averages of the x_i and y_i values. Based on this r-value and the sample size n one can calculate the so-called t-value via

$$t = r_p \frac{\sqrt{n-2}}{\sqrt{1-r_p^2}}. \qquad (6.2)$$

Equation 6.2 can then be converted to a p-value using, for example, a resource such as the p-value from t-score calculator by the Social Science Statistics website.[33]

The Spearman rank correlation coefficient r_s, when there are no tied ranks, can be calculated using the equation

$$r_s = 1 - \frac{6\sum_{i=1}^{n} d_i^2}{n(n^2 - 1)}, \qquad (6.3)$$

where d_i is the difference between ranks for individual i in the sample, with a total sample size of n individuals. It is important to note that the Spearman rank correlation coefficient is only valid for monotonically increasing or decreasing functions, meaning functions that continually increase or decrease but do not do both.

Extreme outliers are data values that are more than three times the interquartile range (IQR) below the first quartile (Q_1) or above the third quartile (Q_3). That is, x is considered an extreme outlier if

$$x < Q_1 - 3(IQR), \text{ or } x > Q_3 + 3(IQR). \qquad (6.4)$$

Q_1 is called the first quartile because 25 percent (or one quarter) of the data lies below it. Similarly, Q_3 is known as the third quartile because 75 percent (or three quarters) of the data lies below it. The IQR defines the middle 50 percent of the data, and hence can be defined via $IQR = Q_3 - Q_1$.

6.8.2 Military Study

The thirty countries used in the study on military spending are:

> China, Saudi Arabia, India, the United Kingdom, Germany, Japan, Russia, South Korea, France, Italy, the Netherlands, Poland, Afghanistan, Pakistan, Singapore, Egypt, Taiwan, Colombia, Morocco, Oman, Romania, Switzerland, Uruguay, Belgium, Greece, Denmark, New Zealand, South Africa, Chile, and Argentina.

Using equation 6.4, we find that an extreme outlier in the x-variable of figure 6.8a is defined as any x-value greater than 0.000057. The number of nuclear detonations per square kilometer in the state of Nevada is 0.0035, which clearly makes it an extreme outlier. This may significantly impact the reliability of the Pearson correlation coefficient in this study. However, the Spearman coefficient should be more reliable as it is less sensitive to extreme outliers. Indeed, the difference between the r_p and r_s values indicate how big an effect the extreme outlier has. We also note that the y-variable for Nevada, namely the number of reports per capita, is not an extreme outlier in this case.

Regarding the study of nuclear warheads (see figure 6.8b), using equation 6.4 we can identify extreme outliers in the x-variable for this data set as those greater than 905 and in the y-variable as those greater than 0.000015. This means that Russia with a massive 6,257 nuclear warheads counts as an extreme outlier, and the UK counts as an outlier in terms of the number of UFO reports per capita. Given the presence of two extreme outliers in an already small sample size, the value for Pearson's correlation coefficient presented in figure 6.8b is likely unreliable. However, the Spearman coefficient presented in figure 6.8b should be more reliable in this case.

6.8.3 Water Study

The distance from the geographical center of the United States along lines of constant latitude, denoted by the variable D_L (see figure 6.9), is modelled using the equation

$$D_L = \sqrt{(L_S - L_{USA})^2}, \tag{6.5}$$

where L_S is the longitude of the geographical center of each US state, measured in degrees from the prime meridian. Likewise, L_{USA} is the longitude of the geographical center of the United States in degrees from the prime meridian. Of course, the definition of equation 6.5 does not define a true physical distance but can easily be converted to a real distance. For example, at the equator one degree of longitude equals about 111 kilometers.

Regarding the study on the coastline length of US states (see figure 6.10), we can identify extreme outliers in the x and y variables using equation 6.4. Using this definition, coastline lengths greater than 19,501 kilometers are considered extreme outliers. Likewise, reports per capita greater than 0.00088 are also extreme outliers. This means that the data point in the upper right corner of figure 6.10 is an extreme outlier. This may affect the reliability of Pearson's correlation coefficient in this study. Therefore, the more reliable measure of correlation in this study is Spearman's rank.

The thirty US states used in the length of coastline study (see figure 6.10) are:

> Alabama, Alaska, California, Connecticut, Delaware, Florida, Georgia, Hawaii, Illinois, Indiana, Louisiana, Maine, Maryland, Massachusetts, Michigan, Minnesota, Mississippi, New Hampshire, New Jersey, New York, North Carolina, Ohio, Oregon, Pennsylvania, Rhode Island, South Carolina, Texas, Virginia, Washington, and Wisconsin.

The twenty-nine countries used in the length of coastline study (see figure 6.11) are:

> Canada, Russia, Indonesia, Chile, Australia, Norway, Philippines, Brazil, Finland, Turkey, Saudi Arabia, France, Spain, Thailand, Mozambique, Venezuela, Ireland, Egypt, Iran, South Africa, Germany, Tanzania, Eritrea, Peru, Bangladesh, Guinea-Bissau, Yemen, Vanuatu, and Nigeria.

The thirty countries used in the study of the fraction of a countries area that is covered in fresh water (figure 6.13) are:

> The United Kingdom, Spain, New Zealand, Brazil, Italy, Canada, Algeria, Greece, Germany, Belgium, Denmark, Austria, Bhutan, North Korea, China, Philippines, Iran, Saudi Arabia, Tunisia, Namibia, Armenia, Guinea, India, Sri Lanka, Lebanon, Ireland, Iceland, Australia, Estonia, and the United Arab Emirates.

6.8.4 Environment Study

The sixteen countries used in the study on spent nuclear fuel are:

> Belgium, Bulgaria, Czech Republic, Finland, France, Germany, Hungary, Lithuania, Netherlands, Romania, Slovenia, Spain, Sweden, Switzerland, Ukraine, and the UK.

Extreme outliers in the x and y variables are identified using equation 6.4. If a particular country exceeds 20,261 tons of spent nuclear fuel it is considered an extreme outlier through this definition. None of the sixteen countries in this study exceed this limit, thankfully for the environment. Equation 6.4 also tells us that if the reports per capita exceed 0.000019 then it is an extreme outlier. Since the UK has 0.000042 reports per capita it is thus considered an extreme outlier. This rogue data point is easily identifiable as the point near the very top of figure 6.15. In this study, it is therefore better to rely on Spearman's rank correlation coefficient rather than Pearson's coefficient.

The thirty countries sampled for the study on CO_2 emissions are:

> The UK, Spain, New Zealand, Brazil, Italy, Canada, Algeria, Greece, Germany, Belgium, Denmark, Austria, Bhutan, North Korea, China, Philippines, Iran, Saudi Arabia, Tunisia, Namibia, Armenia, Guinea, India, Sri Lanka, Lebanon, Ireland, Iceland, Australia, Estonia, and the United Arab Emirates.

6.8.5 Earthquake Study

Regarding figure 6.17, the Pearson correlation coefficient is likely to be significantly affected by the clearly non-normal distribution seen in this data set. This may partly explain the large discrepancy between the r_s and r_p values in this study. In addition, there are extreme outliers in this data set, which further adds to the unreliable value of r_p.

The eighteen countries used in figure 6.18 are:

> Brazil, France, Algeria, Denmark, Spain, Australia, Turkey, Greece, Iran, India, China, New Zealand, Philippines, Japan, Chile, Mexico, Peru, and Indonesia.

6.8.6 Blood Study

Values for the percentage of rhesus negative blood range from as low as 0.3 percent up to as high as 19 percent. The number of UFO reports made to NUFORC per capita range from 0 up to a maximum of 0.000020 for Australia. Using equation 6.4, we find that an extreme outlier is defined as any country with a rhesus negative blood percentage greater than 35.8 percent, or any country with greater than 0.000023 reports per capita. Therefore, we can conclude that this data set contains no extreme outliers as defined by equation 6.4.

The fifty-one countries used in this study (see figure 6.19) are:

> Germany, Austria, North Korea, Philippines, Saudi Arabia, Tunisia, Namibia, Armenia, Guinea, India, Sri Lanka, Ireland, Australia, Estonia, United Arab

Emirates, Bolivia, Jordan, Cambodia, Gabon, Peru, Uzbekistan, Malta, Estonia, Morocco, Argentina, Azerbaijan, Portugal, Lithuania, Somalia, the Netherlands, Kazakhstan, Ecuador, Honduras, Nigeria, Libya, Burkina Faso, Costa Rica, Jamaica, Sudan, Finland, Romania, Ukraine, Serbia, Slovenia, Sweden, Jordan, Israel, Pakistan, Turkey, Colombia, and Algeria.

These countries are sampled from around the world and display an approximately normal probability distribution ($P(x)$) in the percentage of rhesus negative blood (x), as shown in figure 6.20.

The data presented in figure 6.20 appears to be approximately normally distributed, although a Pearson χ^2 test for normality gives a test statistic of 3.3 and p-value of 0.85, indicating the fit is not great. The other variable, namely the number of reports per capita, does not approximate a normal distribution at all. However, as per the discussion in the literature this may not affect the reliability of the Pearson correlation coefficient result. In any case, the Spearman rank correlation coefficient, which does not require a normal distribution, supports the conclusion of a statistically significant large positive correlation.

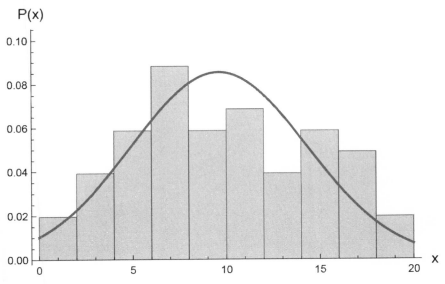

FIGURE 6.20
The discrete probability distribution for the percentage of rhesus negative blood among the fifty-one countries sampled. Data is grouped in bin sizes of 2 percent and fitted to a continuous normal distribution (continuous curve). The data appears approximately normally distributed.

Looking at figure 6.19, the data seems to become slightly more spread (bigger variance) as we move along the trend line to greater values of the variables. Thus, the data does not display perfect homoscedasticity. The data in figure 6.19 is at least approximately linear.

Overall, the conclusion that a statistically significant large positive correlation exists between the percentage of a country's population with rhesus negative blood and the number of UFO sightings reported to NUFORC appears to be robust. However, further study needs to be conducted to definitively establish the validity, and possible causation, of this result.

6.9 Notes

1. Sightings, *Sightings Russian Cosmonaut UFO Sightings*, 1992. (accessed May 10, 2022, https://www.youtube.com/watch?v=N0S4M_aFYCM).

2. R. Proctor, "The History of the Discovery of the Cigarette–Lung Cancer Link: Evidentiary Traditions, Corporate Denial, Global Toll," *Tobacco Control* 21 (2012): 87–91.

3. June Liu,. "Correlation Among Different Variables and Life Expectancy," *Undergraduate Journal of Mathematical Modelling: One + Two* 3, no. 2 (2011): article 2.

4. Bland, *Clinical Biostatistics: The Correlation Coefficient*, 2006 (accessed May 24, 2022, https://www-users.york.ac.uk/~mb55/msc/clinbio/week7/corr.htm).

5. Amitav Sarma et al., 2020. "Correlation between the arm-span and the standing height among males and females of the Khasi tribal population of Meghalaya state of North-Eastern India." *Journal of family medicine and primary care* (2020): 6125–29 (https://www.ncbi.nlm.nih.gov/pmc/articles/PMC7928122/).

6. S. E. Edgell and S. M. Noon, "Effect of Violation of Normality on the Test of the Correlation Coefficient," *Psychological Bulletin* 95, no. 3 (1084): 576–83.

7. Robert V. Hogg and Allen T. Craig, *Introduction to Mathematical Statistics* (New York: Pearson, 2014).

8. NUFORC. (accessed May 2022, https://www.nuforc.org/webreports/ndxevent.html).

9. NICAP, *NICAP*, n.d. (accessed May 2022, https://nicap.org/books/aadkw/aadkw_ch3.htm).

10. Lee Speigel, "UFOs Confront Soldiers During Way," *Huffington Post*, April 19, 2015.

11. John Greenewald Jr., "theblackvault," n.d. (accessed May 2022, http://documents.theblackvault.com/documents/ufos/malmstromufo.pdf).

12. Robert L. Hastings, *UFOs and Nukes: Extraordinary Encounters at Nuclear Weapons Sites* (Scotts Valley, CA: CreateSpace Independent Publishing, 2008).

13. NASA, *Refueling in Space to Help Enable Exploration*, n.d. (accessed May 24, 2022, https://www.nasa.gov/offices/education/centers/kennedy/home/OASIS.html).

14. RCMP, "National archive of Canada (file number RCMP 67-400-23-X.jpg)," n.d. (accessed May 10, 2022).

15. BBC, *Zimbabwe's Mass UFO Sightings*, June 21, 2021 (accessed May 2022, www.bbc.co.uk/programmes/w3ct1x0z).
16. *The Phenomenon* (2020). Directed by James Fox.
17. T. Laurent et al., "Spatial Point Pattern Analysis of the Unidentified Aerial Phenomena in France," *arXiv.org*, 2015.
18. CRS, "Nuclear Waste Storage Sites in the United States," 2020 (accessed May 2022, https://sgp.fas.org/crs/nuke/IF11201.pdf).
19. World Nuclear Waste (WNW), "The World Nuclear Waste Report 2019," *boell*, 2019 (accessed May 2022, https://www.boell.de/sites/default/files/2019-11/World_Nuclear_Waste_Report_2019_Focus_Europe_0.pdf).
20. WNW, "The World Nuclear Waste Report."
21. Richard Clarke and R. P. Eddy, *Warnings: Finding Cassandras to Stop Catastrophes* (New York: Harper Collins, 2018).
22. M. Persinger and J. Derr, "Geophysical Variables and Behavior: XXIII. Relations between UFO Reports within the Uinta Basin and Local Seismicity," *Percept Mot Skills* 60, no. 1 (1985):143–52.
23. Persinger, "Geophysical Variables."
24. Ibid.
25. G. Anagnostopoulos, "A Study of Correlation between Seismicity and Mental Health: Crete, 2008–2010," *Geomatics, Natural Hazards and Risk* (2013): 45–75.
26. Columbia University, *Globally, Only Half of Women Get Treatment for Rh Disease*, August 2020 (accessed May 25, 2022, https://www.cuimc.columbia.edu/news/globally-only-half-women-get-treatment-preventable-killer-newborns#:~:text=A%20treatment%20developed%20over%2050,Columbia%20University%20Irving%20Medical%20Center).
27. Young Kim et al., "Relationship between Blood Type and Outcomes Following COVID-19 Infection," *Elsevier Public Health Emergency Collection* (2021): 125–31.
28. I. Mouro, Y. Colin, B. Chérif-Zahar, J. P. Cartron, C. Le Van Kim, "Molecular Genetic Basis of the Human Rhesus Blood Group System," *Nature Genetics* 5, no. 1 (1993): 62–65. doi: 10.1038/ng0993-62.
29. Dr. Barry Starr, . "Did Rh- blood come from Neanderthals?" *The Tech Interactive*, July 9, 2013 (accessed May 2022, https://www.thetech.org/ask-a-geneticist/rh-did-not-come-neanderthals).
30. B. Carritt, T. J. Kemp, and M. Poulter, "Evolution of the Human RH (Rhesus) Blood Group Genes: A 50 Year Old Prediction (Partially) Fulfilled," *Human Molecular Genetics* 6, no. 6 (1997): 843–50. doi: 10.1093/hmg/6.6.843.
31. "Blood type distribution by country," *Wikipedia* (accessed January 2022, https://en.wikipedia.org/wiki/Blood_type_distribution_by_country).
32. Flegr, J. Flegr, R. Kuba, and R. Kopecký, "Rhesus-Minus Phenotype as a Predictor of Sexual Desire and Behavior, Wellbeing, Mental Health, and Fecundity," *PLoS ONE* 15, no. 7 (2020): e0236134.
33. "P Value from T Score Calculator," *Social Science Statistics* (accessed January 2022, https://www.socscistatistics.com/pvalues/tdistribution.aspx).

7

When Are All the UFOs?

[In response to a government request that the FBI investigate UFOs.] I would do it, but before agreeing to do it, we must insist upon full access to discs recovered. For instance, in the LA case, the Army grabbed it and would not let us have it for cursory examination.

—J. Edgar Hoover (FBI director)[1]

7.1 Introduction

ON NOVEMBER 5, 1975, a logging crew of seven were cutting and piling trees in Apache-Sitgreaves National Forest near Snowflake, Arizona. After a long day of manual labor, the crew climbed in the truck and made for home along the old logging roads. Someway down the bumpy road a light was spotted shining through the pine and fir trees. By this time the sun had set. As they drove closer, the source of the light became clearer; a golden disc-shaped object hovering silently twenty feet above the ground. One of the crew, Travis Walton, got out of the truck and approached the luminous object. A blue-green ray shot from the craft and hit Travis on the head and chest. He was hurled ten feet through the air, before hitting the floor and remaining lifeless. The crew panicked and hastily drove off.[2]

Travis woke up on a raised table below a bright light. His shirt and jacket were bunched up to his shoulders. There appeared to be a medical device around his abdomen. He rolled his head to the side—and saw three figures. This is how he described them:

> The beings were a little under 5 feet in height. . . . They had a basic humanoid form: two legs, two arms, hands with five digits each, and a head with the normal human arrangement of features. . . . They were thin, puny, covered with marshmallow looking skin. Their small hands were delicate and without nails. Totally bald, their heads were disproportionately large for their little bodies.
> The only facial features that didn't appear underdeveloped were those incredible eyes! Those glistening orbs had brown irises twice the size of those of a normal human's eye, nearly an inch in diameter! The iris was so large that even parts of the pupils were hidden by the lids, giving the eyes a certain catlike appearance. There was very little of the white part of the eye showing. They had no lashes and no eyebrows. Their little mouths never moved.[3]

It is not known whether this account is true. What is known is that despite a thorough search conducted in the area by some fifty volunteers equipped with dogs and helicopters, Travis could not be located for five whole days. The police began suspecting he had been murdered by his co-workers. Eventually, he turned up one night on a road west of Heber, Arizona, some thirty miles from where he was last seen. All six of his co-workers that witnessed the beam of light strike Travis took polygraph tests, and all passed, except one that was inconclusive. Wood core samples taken from trees at the site indicate unusually rapid growth for fifteen years following the incident.[4] Soil samples taken from the site contain interesting anomalies, including unusual quantities of iron.[5]

In this chapter, we focus on one aspect of Travis's experience, his vivid description of the humanoids surrounding him. Are there any clues in this description that may suggest where, or indeed when, these entities came from? It's a well-established fact that humans evolved from apelike ancestors over a period of approximately six million years. But the process of evolution did not just magically stop when it reached modern humans. We are still evolving. We may then ask: *Given another six million years of evolution, what might we look like?* The short answer is that we don't know for sure, but we can extrapolate current trends and make an educated guess.

Paedomorphosis is the retention of juvenile traits into later life. Over the long course of human evolution, there has been a general and gradual trend toward paedomorphosis, a trend that rapidly accelerated over the last 80,000 years.[6] Put simply, adult humans now look much more infantile than they did over 80,000 years ago. This evolutionary trend has resulted in adult humans having larger eyes, bigger heads, less facial hair, and smaller faces. More baby-like. Assuming these trends continue into our distant future, our facial-cranial features would begin to resemble, somewhat eerily, the description provided by Travis Walton and other alleged abductees.

The course of evolution is also shaped by the environment a species evolves in. What are the most likely environmental factors in our far future? One possibility is our continuing exploration of space. Muscular atrophy and bone density depletion are real problems even for today's astronauts, due to low or zero gravity, a problem that will only become greater as we spend longer in space. Thus, we might expect a space-faring human species to gradually evolve reduced muscularity and bone density. Lower light in the darkness of space may also result in us evolving larger pupil sizes to collect more light. Once again leading to similarities with the quintessential depiction of extraterrestrials as being thin, skinny, and with large eyes.

There are even remarkable similarities with present-day humans. Anatomically, the beings are typically described as a little under five feet in height and having a basic humanoid form, with two arms, two legs, and a head. Each hand having five digits. Crucially, they are also described as bipedal, standing upright on two legs. Assuming they evolved independently of humans, such anatomical similarities are extraordinarily unlikely according to Dr. Michael P. Masters, professor of biological anthropology at Montana Technological University. Indeed, as written in his thoroughly researched and excellent book, *Identified Flying Objects: A Multidisciplinary Scientific Approach to the UFO Phenomenon*, Masters writes

> In fact, I would argue that the probability of a fleshy, big-brained, bipedal, pentadactyl [having five fingers or toes], highly intelligent lifeform arising independently on a different planet orbiting a nearby star is effectively zero, given the countless factors necessary to create and sustain this same evolutionary trajectory over such an extraordinary amount of time.[7]

So, assuming the accounts of these beings are real and given the vanishingly small probability that they could have evolved such physiological similarities to us, then we are left with an interesting possibility. Perhaps they look so much like us because, dare I say it, they are us.

As preposterous as this sounds, it does tie together several disparate aspects of the phenomenon, in addition to the already discussed physiological similarity between us and them. A common aspect of UFO sightings is their spontaneous appearance and disappearance. This puzzling observation may also be explained if UFO craft were essentially time machines, since they would appear at a specific point in time without having traversed the intermediate distance in space. Also, if they really are us from the future, then the motivation behind reported warnings over the state of the environment start to make more sense. The time-traveling beings would desperately want to ensure the planet remains habitable, since they would ultimately inherit it, and would therefore feel compelled to warn against the existential dangers of climate change, nuclear weapons, and pollution. Does the abduction phenomenon also fit with this idea? Estimates vary from several thousand alleged abductees worldwide up to 3.7 million in America alone.[8] If this widespread experience is due to extraterrestrials, then why on Earth would they travel across the vastness of space to medically examine us? Well, imagine an archaeologist six million years from now. What better way to study your past than to visit it, observe it, even examine it? To quote professor Masters:

> We would no longer be left to scrape through layers of dirt to discover our origins and evolutionary history, as we could instead conduct cross-temporal . . . biomedical examinations of our hominin ancestors, gaining a much clearer understanding of their biology, morphology, and changing patterns of culture and cognition. It is this aspect of abduction reports that are most intriguing, as the processes described are precisely what we anthropologists would do in order to learn as much about the human past as possible. . . . Everything about these visits, as described by those who claim to have been a part of them, seem to indicate that the principal aim is to collect data about the ancestral human past.[9]

This temporal hypothesis also comes with a lot of problems, which mainly stem from our current understanding of physics. Our best description of gravity is presently defined by Einstein's beautiful theory of general relativity. Deeply hidden within the complex mathematics of Einstein's theory are a few rather bizarre solutions, solutions that allow strange trajectories through space and time known as closed time-like curves (CTCs). CTCs seem to permit time travel into the past, at least theoretically. By following a closed time-like curve, an intrepid explorer could journey into the future but arrive in the past. This raises potentially insurmountable logical contradictions that come from interacting with one's own past. For example, imagine going back in time and murdering your grandfather before he conceived your father or mother. If that were possible, you could never have been born in the first place and would therefore not be able to do anything, let alone travel back in time and murder someone. A logical contradiction.

However, many physicists have argued that CTCs may not in fact lead to logical contradictions after all. One idea, known as the Novikov self-consistency principle, conjectures that if some event (for example, killing your grandfather) could cause a logical paradox then the probability of that event occurring is precisely zero. According to this principle, the laws of physics would somehow conspire to prevent you from killing your grandfather, thus maintaining self-consistency. How this would happen in practice is unknown. Another possible resolution may be that any change made to the past is corrected for so long that eventually it's like the change never occurred in the first place. For example, as the authors of the peer-reviewed article "Reversible Dynamics with Closed Time-like Curves and Freedom of

Choice" published in the prestigious journal *Classical and Quantum Gravity* said, "You might try and stop patient zero from becoming infected [with coronavirus], but in doing so, you would catch the virus and become patient zero, or someone else would."[10] In this scenario, some of the insignificant details might change, such as the name of patient zero, but the salient features would not. Millions of people would still die. And we would still all have to sit through Zoom meetings in pajama bottoms while listening to background noise from someone that did not mute their microphone.

But the truth is, we simply don't know enough physics to know if time travel to the past is possible. Our current understanding of the universe hinges on two theories: Einstein's theory of general relativity and quantum mechanics. These two theories work remarkably well on their own, however, there are some physical scenarios, such as in the very early universe or near black holes that can only be described by both theories at the same time. In these specific scenarios, we are completely lost. Physically measurable quantities that should have nicely behaved and predictable values become nonsensical, often blowing up to infinite values. The two theories appear completely incompatible, defying all attempts at reconciliation, at least so far. We just don't know enough physics to even contemplate time travel to the past, although this may change if we do one day manage to reconcile general relativity and quantum mechanics into a so-called theory of quantum gravity.

In any case, the hypothesis that *they* are just us in the far future is testable. As proposed by Professor Masters, we can test this hypothesis by simply waiting thousands or millions of years to see if we evolve into the entities described and develop the technology reported. Not the quickest way to test a hypothesis. A faster, but less direct, way to examine this hypothesis is to look for statistically significant spikes in UFO reports during key events in history. Imagine that scientists suddenly announce they have invented a time machine—and offer a free ride. Most people would want to see specific times or events like the construction of the pyramids in Egypt, the destruction of the library of Alexandria, or the sealing of Magna Carta. Unfortunately, we do not have reliable records of UFO sightings that go back this far in time, but we do have a database of UFO reports from 1940 to the present day. Is there a spike in UFO reports during certain historical events like 9/11 or the start of the coronavirus pandemic? On the other hand, perhaps any such spikes in report numbers could have entirely innocent and prosaic explanations such as increased reporting due to the spread of the internet, the release of popular UFO movies, or even meteor showers? Let's analyze the data to find out.

7.2 Analysis

Was there an unusual amount of UFO activity around the time of the terrorist attacks on the World Trade Center in September 2001? To answer this question, we first need to know the normal level of UFO activity for a typical September. We look at the number of reports made in the month of September for five years before and five years after 2001. Averaging over September report numbers for the years 1996 to 2006 will then give us a baseline against which we can compare the September 2001 numbers. Referring to table 7.1, the mean average number of reports filed to NUFORC in the month of September for the years 1996 to 2006 is found to be 330. The sample standard deviation for this data set is 153.4 reports (see appendix 3.3.1 for the definition of standard deviation). Therefore, the number of UFO reports in September 2001 (356) was greater than the average over the period 1996 to 2006 (330), but not by a lot. The September 2001 number is only 0.2 standard deviations above the average. Not a statistically significant increase.

TABLE 7.1
A table of the number of UFO sightings reported to NUFORC in the month of September for the years 1996 through 2006.

Year	September reports
1996	77
1997	60
1998	251
1999	407
2000	301
2001	356
2002	384
2003	512
2004	417
2005	533
2006	332

Okay, so that's for the entire month of September 2001. But what about individual days within this month? Perhaps a surge in UFO cases will be more apparent on a day-to-day basis? Like the method we just used, we compare the number of reports on September 11, 2001, with an average over the five days before and after this infamous date. The average number of reports per day over the period September 6 through 16, 2001, is 17.1. The sample standard deviation for this data set is 10.4 reports. Therefore, the number of UFO reports on September 11, 2001, was greater than the average over the period September 6 through 16, 2001, by a substantial 2.6 standard deviations. This is an interesting result since it is above the 2 σ threshold.

However, we must be cautious and think carefully about possible prosaic causes for this result. For example, 9/11 was a frightening event, the average person would likely have been more anxious and paranoid than usual. A greater number of people would have been looking to the September sky, fearfully searching for more threats. Also, some of the reports to NUFORC were filed retrospectively, in some cases months or years after the observed UFO. If a witness could not exactly recall the date, I suspect a momentous historical date such as 9/11 could accidentally take its place. Whatever the reasons, it is still very interesting that there was such a significant spike in UFO sightings on the specific date of September 11, 2001.

TABLE 7.2
A table of the number of UFO sightings reported to NUFORC from September 6 to 16 of 2001.

Day in September 2001	Reports
6	16
7	5
8	9
9	12
10	16
11	44
12	18
13	20
14	8
15	21
16	19

Another infamous event in our recent history was the start of the global coronavirus pandemic. Patient zero is the first documented patient in a disease epidemic or pandemic. They are generally considered the origin of all future cases, the epicenter if you will. The identity of coronavirus patient zero is not known for sure. However, one possible candidate was a fifty-five-year-old from Hubei province in China, who may have contracted COVID-19 on November 17, 2019.[11] Regardless of who patient zero was, it seems likely that the first human infection took place sometime in November 2019.[a] Were there a bigger than normal number of UFO sightings in November 2019?

---

a. Although there is some indirect evidence that suggests it could have been slightly earlier than this date.

In November 2017, a total of 365 UFO reports were submitted to NUFORC. In November 2018, this dropped to 259, before rising to a peak of 702 in November 2019. In November 2020 and 2021, the numbers of reports declined to 452 and 252, respectively. The average number of reports during the month of November for the years 2017 to 2021 is therefore 406. The sample standard deviation is 184.9 reports. The number of reports for November 2019 is therefore 1.6 standard deviations above the average for the period 2017 to 2021. This result is not above the 2σ threshold, although it is close. Unfortunately, since the precise day in November 2019 of the first coronavirus infection is unknown, we cannot do an analysis on a day-to-day timeframe.

We now zoom out and consider a much bigger timeframe. In figure 7.1a, we plot the number of worldwide UFO reports over an eighty-one-year period, namely from 1940 to 2021. On this scale, it appears there were very few reports from the early 1940s all the way up until the mid-1960s, when the first significant upsurge began. This was followed by another temporary increase in reports around the mid-1970s. Then nothing much for the next twenty years, with a fairly constant number of reports between approximately 1975 and 1995. However, beginning around the mid-1990s, there was a dramatic surge in reports that lasted until around 2014, as can be seen from figure 7.1a. One possible explanation for this specific surge in UFO reports could be the simultaneous explosion in the number of internet users from the mid-1990s onward. However, bizarrely, and somewhat against the pattern established over the previous seventy-five years or so, a substantial decline in reports began around 2014 and continues to the present.

Regardless of these local fluctuations in report numbers, the average has clearly increased over time. It is also clear from figure 7.1a that this increase is non-linear, meaning that a curved trend line fits the data better than a straight line. In fact, we find that the number of reports as a function of time can be modelled reasonably well by an exponentially increasing function, as shown by the grey band in figure 7.1a (see appendix 7.3.1 for more details). If the number of reports really is exponentially increasing with time, then taking the logarithm of the number of reports should yield data that closely follows a straight line of best fit.[b] The logarithm of the number of reports worldwide from 1940 to 2021 can be seen in figure 7.1b. On this logarithmic scale, the data do indeed appear to fluctuate quite evenly about the straight line of best fit, indicating the exponential model is reasonable.

The logarithmic scale used in figure 7.1b has another principal advantage. Periodic surges in report numbers can be more readily identified as signifi-

---

b. The mathematical reason for this can be found in appendix 7.3.1.

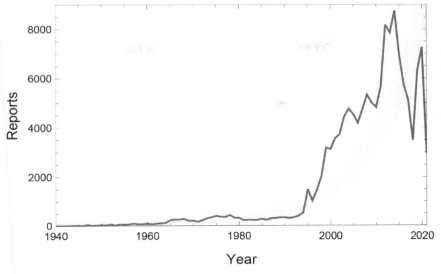

FIGURE 7.1a
(a) The number of reported UFO sightings worldwide from 1940 to 2021. The light grey band shows an exponential fit to the data (see appendix 7.3.1). (b) The natural logarithm of the number of reported UFO sightings worldwide from 1940 to 2021, including a straight line of best fit. 1947, 1951, 1957, 1967, 1975, and 2006 are highlighted by dashed gray lines, signifying the mid-point of each temporary surge in report numbers.

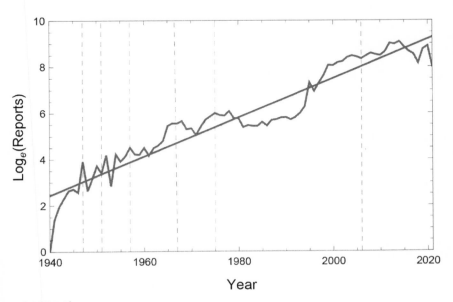

FIGURE 7.1b

cant fluctuations above the straight line of best fit. From this plot, the years 1947, 1951, 1957, 1967, 1975, and 2006 are identified as the centers of particularly significant periods of UFO activity. These key years are indicated by dashed grey vertical lines in figure 7.1b.

Do these localized surges in UFO report numbers coincide with the dates of conflicts around the globe? To address this question, we look at six of the most significant wars between 1940 and 2021, namely: (US involvement in) the Second World War (1941–1945), the Korean War (1950–1953), the Vietnam War (1961–1973), the Gulf War (1991), the Iraq War (2003–2010), and the war in Afghanistan (2001–2021). The middle year for each of these conflicts, rounded to the nearest whole year, is 1943, 1952, 1967, 1991, 2007, and 2011, respectively. Comparing these dates one-for-one with those corresponding to surges in UFO reports yields an average difference of approximately thirteen years. Far from synchronous. Even if we hand-pick the four wartime dates that most closely match surges in UFO reports, namely 1943, 1952, 1967, and 2007, the average difference is still off by around two years. Not particularly coincidental.

Furthermore, the timing with respect to nuclear related events is not particularly convincing either. For example, the world's first man-made nuclear chain reaction occurred in December 1942. The first test of a nuclear weapon occurred in July 1945, with the Trinity test in New Mexico. This was soon followed by the first use of nuclear weapons in warfare in August 1945. All these dates took place at least two years before the first surge in UFO reports in 1947.

Perhaps there could be a far more prosaic explanation for the periodic surges seen in UFO reports. It is hard to disentangle UFOs from pop culture, and pop culture from UFOs. Our fascination with extraterrestrials has led to a substantial market for UFO movies, TV shows, books, and documentaries. But perhaps the reverse is also true? It is difficult to overestimate the cultural impact of TV shows like *The X-Files*, or movies like *Close Encounters of the Third Kind* or *E.T. the Extra-Terrestrial*. It seems plausible that such culturally significant works would influence people's perception of unidentified objects in the sky. Are we more likely to mistake something ordinary, such as a shooting star or airplane, for something extraordinary like an alien spacecraft after watching a movie about extraterrestrials? Does art imitate reality, or reality imitate art?

We can hopefully gain insight into this question using statistics. In figure 7.2a, we plot the number of movie releases per year featuring extraterrestrials between 1940 and 2021.[12] Using this data, we identify seven localized spikes in the number of E.T. movie releases (1953, 1958, 1966, 1984, 1999, 2011, and 2019), as shown by the dashed gray lines in figure 7.2a. Comparing the first six of these dates with the dates for surges in UFO reports yields an average difference of eleven years. Not such a close match.

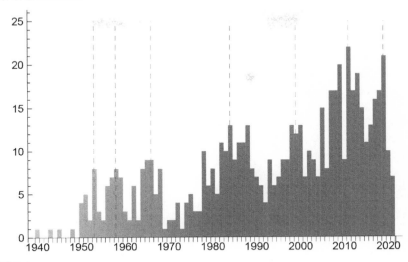

FIGURE 7.2a
(a) The number of movies featuring extraterrestrials released each year between 1940 and 2021. The gray dashed lines indicate years with a local spike in the number of E.T. movie releases. (b) The number of movies featuring extraterrestrials released each year against the number of reported UFO sightings worldwide each year between 1950 and 2021. The r-values indicate a large positive correlation, and the p-values show statistical significance at the 1 percent level.

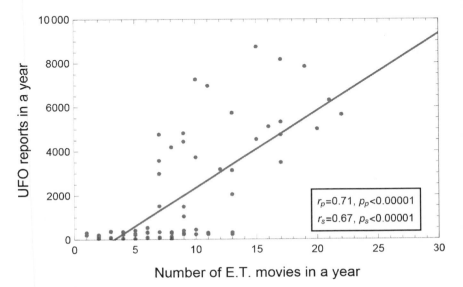

FIGURE 7.2b

However, it is possible to analyze this relationship in more detail, by looking for a correlation on a yearly basis. In figure 7.2b, the number of movie releases featuring extraterrestrials is plotted against the number of reported UFO sightings worldwide for every year between 1950 and 2021. We change to a start year of 1950 because there were essentially no E.T. movie releases between 1940 and 1949, as can be seen in figure 7.2a. For the Pearson test, we observe a statistically significant large positive correlation ($r_p = 0.71$, $p_p < 0.00001$).[13] Results from a Spearman rank test corroborate this finding ($r_s = 0.67$, $p_s < 0.00001$) by also establishing a statistically significant large positive correlation.

Taking this result at face value suggests that E.T. movie releases could explain at least some proportion of UFO sightings. However, once again, we must caution that correlation does not necessarily equal causation. In fact, the causal relationship could be the other way around. It could be that a greater number of UFO sightings leads to a greater demand and revenue for UFO movies. Moreover, I would again emphasize that popular culture may explain some UFO sightings, but I find it unlikely that it could account for all such observations. Primarily, this is because many sightings are also independently corroborated by sensor data, such as radar returns. But also, because UFO sightings frequently occur in almost every country across the globe and in every culture, even those with little or no widespread access to popular culture mediums such as television, cinema, and the internet.

What time of the year are you most likely to see a UFO? A popular hypothesis is that UFOs are spotted more often in the summer months, because people spend more time outdoors enjoying the nice weather. We can test this hypothesis by simply counting the total number of UFO reports each month over a period of many years. The results of this are presented in figure 7.3a for the United States and figure 7.3b for Canada. For both countries, July has the most UFO sightings. So, this verifies the hypothesis, right? Well, maybe not.

What it verifies is that UFOs are reported more often in July, but this does not necessarily equate to them being reported more often in the summer. This is because July is only a summer month in the northern hemisphere. In the southern hemisphere July is in mid-winter. A seemingly benign distinction. But a distinction that allows us to test whether one is more likely to see a UFO in the summer, and whether this is due to increased time spent outdoors. All we need do is repeat the above process for two countries in the southern hemisphere. This is exactly what is done in figure 7.3c for Australia and figure 7.3d for Brazil.

As can be seen in figure 7.3c and figure 7.3d, June has the most UFO sightings for both southern hemisphere countries. Therefore, it does not seem to be the season that is the most important factor, but rather the month of the

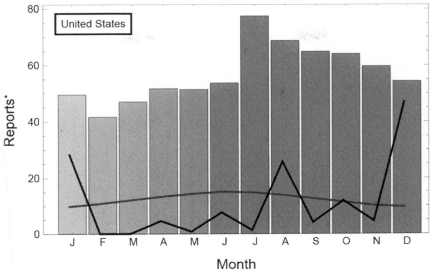

FIGURE 7.3a
Total UFO reports per month in (a) The United States from 2011 to 2021 (reports divided by 100); (b) Canada from 1930 to 2021; (c) Australia from 1930 to 2021; and (d) Brazil from 1970 to 2021. The smoother curves show the average number of daylight hours per month for each country. The jagged curves show the rescaled meteor frequency per month.

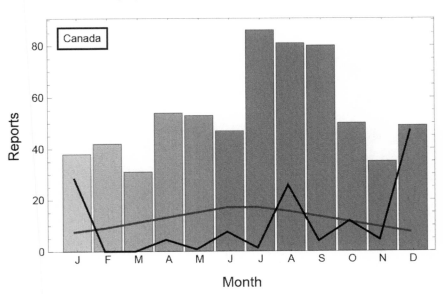

FIGURE 7.3b

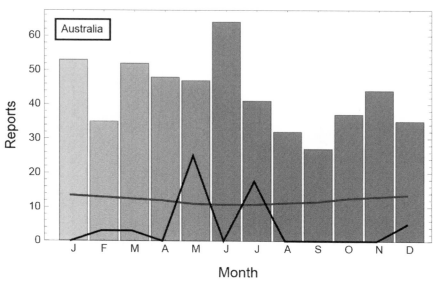

FIGURE 7.3c

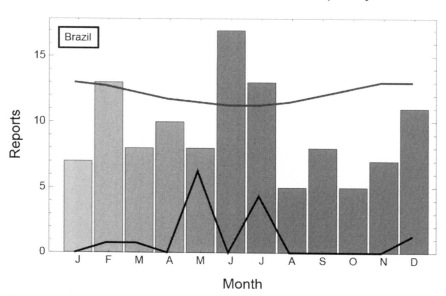

FIGURE 7.3d

year. All four graphs presented in figures 7.3a to 7.3d contain a smooth curve showing the average number of daylight hours per month for each country. For the northern hemisphere countries, the greatest number of daylight hours are in June and July. Yet, for Australia and Brazil in the southern hemisphere, the smallest number of daylight hours are in June and July. Assuming more daylight hours equates to more time spent outdoors, as one would naturally assume, the hypothesis that UFOs are spotted more often in the summer months because people spend more time outdoors seems unlikely based on this analysis.

Then why June and July? One possibility is due to the Earth's position in its orbit around the sun during these months. Most people think the Earth is closest to the sun in July, but it is closest in January. Could the increase in UFO sightings in June and July somehow be due to a larger distance between the sun and the Earth? The Earth is furthest from the sun (aphelion) in early July each year, thus very close to the June-July boundary. Due to Earth's aphelion in July, it's possible that some unknown electromagnetic interaction between charged particles ejected from the Sun (solar wind) and the Earth's magnetosphere produces light displays that are mistaken for UFOs. However, this seems unlikely given the predictably periodic nature of longer-term solar cycles, and their mismatch with surges in UFO reports.

Meteor showers occur at regular intervals as the Earth passes through the tail of dusty debris left by a comet as it journeys around the sun. One example comes from perhaps the most famous meteor shower, the Perseids, which peak in August every year. The Perseid meteor shower occurs because the Earth passes through the debris field left from the comet Swift-Tuttle, which predictably swings by the sun every 135 years. Could meteor showers explain the excess UFO sightings in June and July?

In figure 7.3, the jagged lines show the meteor frequency per month as seen in the northern hemisphere (figures 7.3a and 7.3b) and southern hemisphere (figures 7.3c and 7.3d). The meteor frequency is measured in units of the zenithal hourly rate (ZHR), and the values have been rescaled to better fit on the graph (see appendix 7.3.3 for details). ZHR is a measure of the number of meteors an observer would see in an hour during peak activity, assuming perfect viewing conditions. In the northern hemisphere the meteor frequency is highest in December and January, which clearly does not coincide with the peak months of June and July for UFO reports. While in the southern hemisphere, meteor frequency peaks in May and July, which vaguely coincides with peak UFO sightings. Therefore, it seems unlikely that meteor showers can adequately account for all the excess UFO reports made in June and July, although it is possibly a contributing factor in the southern hemisphere.

7.3 Appendix

7.3.1 Calculations

The exponential fit shown in light grey in figure 7.1a is of the form

$$y = ae^{mx}, \tag{7.1}$$

where a and m are constants. Taking the natural logarithm (logarithm of base e, where e ≈ 2.718 is Euler's number) of both sides of equation 7.1 yields

$$log_e(y) = log_e(ae^{mx}) = log_e(a) + log_e(e^{mx}) = mx + log_e(a), \tag{7.2}$$

where we have used the logarithmic identities

$$log_b(xy) = log_b(x) + log_b(y), log_b(b^{mx}) = mx. \tag{7.3}$$

If we now let $Y = log_e(y)$ and $c = log_e(a)$ then we obtain the result

$$Y = mx + c, \tag{7.4}$$

which is the equation of a straight line with gradient m and y-intercept c.

7.3.2 E.T. Movies Study

Figure 7.2b does not appear to contain any obvious outliers. Let's check whether this is really the case by applying equation 6.4 to this data set. This definition tells us that any year exceeding twenty-nine E.T. movie releases define an extreme outlier. Likewise, any year with more than 13,930 UFO reports also defines an outlier. Based on these limits, figure 7.2b contains no extreme outliers.

Next, we test whether our data set is approximately normally distributed. The results are presented in figure 7.4. The data is grouped in bin sizes of two and fitted to a continuous normal distribution. The data appears approximately normally distributed. However, Pearson's χ^2 test for goodness-of-fit yields a test statistic of 12.333 and a p-value of 0.195, indicating the fit to a normal distribution is far from perfect. The data in figure 7.4 also appears to be heavily skewed towards lower values of the x-variable. These facts indicate that the Pearson correlation coefficient is likely to be unreliable in this study. The Spearman rank coefficient should therefore be considered the more reliable of the two measures.

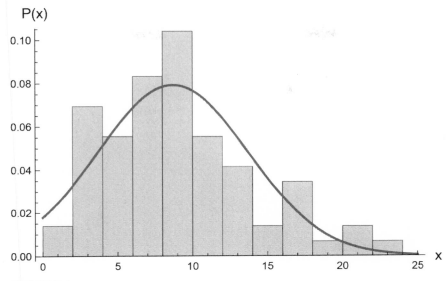

FIGURE 7.4
The discrete probability distribution for the number of E.T. movies each year between 1950 and 2021.

7.3.3 Time of Year Study

In this study, the month in which a meteor shower occurs is defined by the month in which the day of peak activity occurs.[14] Meteor showers with a positive declination are taken to be predominately northern hemisphere events, and those with negative declinations are taken to be predominately southern hemisphere events. A declination of zero degrees is placed in the northern hemisphere category.

The hours of sunlight for the United States, Canada, Australia, and Brazil are determined via meteorological data specific to: Washington, DC, Alberta, Alice Springs, and Brasilia, respectively.

7.4 Notes

1. Federal Bureau of Investigation (FBI), *The Official FBI Reports on Unidentified Flying Objects* (Cabin John, MD: Wildside Press, 2009).

2. Travis Walton, *The Aliens*, n.d. (accessed May 25, 2022, https://www.travis-walton.com/aliens.html).

3. B. J. Booth, *The Travis Walton Abduction, Part 2*, n.d. (accessed May 25, 2022, https://www.ufocasebook.com/Walton.html).

4. John Greenewald, *theblackvault*, n.d. (accessed May 25, 2022, https://documents.theblackvault.com/documents/Budinger/UT095.pdf).

5. Greenewald, *theblackvault*.

6. Dr. Michael P. Masters, *Identified Flying Objects: A Multidisciplinary Scientific Approach to the UFO Phenomenon* (Pennsylvania, PA: Masters Creative LLC, 2019).

7. Masters, *Identified Flying Objects*.

8. Dr. Ema Sullivan-Bissett. n.d. "Is it Normal to Believe You Have Been Abducted by Aliens?" n.d. (accessed May 26, 2022). https://www.birmingham.ac.uk/research/perspective/abducted-by-aliens.aspx.

9. Masters, *Identified Flying Objects*.

10. Germain Tobar and Fabio Costa, "Reversible Dynamics with Closed Time-like Curves and Freedom of Choice," *Classical and Quantum Gravity* 37, no. 20 (2020): 205011.

11. Josephine Ma, "Coronavirus: China's First Confirmed COVID-19 Case Traced Back to November 17," *South China Morning Post*, March 13, 2020 (accessed May 26, 2022), https://www.scmp.com/news/china/society/article/3074991/coronavirus-chinas-first-confirmed-covid-19-case-traced-back).

12. "List of films featuring extraterrestrials," *Wikipedia* (accessed May 26, 2022, https://en.wikipedia.org/wiki/List_of_films_featuring_extraterrestrials).

13. Tosin Adekanye, "An Analysis of UFO Sightings," *TosinLitics*, August 26, 2021 (accessed May 26, 2022, https://medium.com/low-code-for-advanced-data-science/an-analysis-of-ufo-sightings-f261c08e9dc3).

14. The data used in this study can be found in "List of meteor showers," *Wikipedia* (accessed March 1, 2022, https://en.wikipedia.org/wiki/List_of_meteor_showers).

8

What Does It All Mean?

I believe that these extraterrestrial vehicles and their crews are visiting this planet from other planets, which are a little more technically advanced than we are on Earth. I feel that we need to have a top level, coordinated program to scientifically collect and analyze data from all over the Earth concerning any type of encounter, and to determine how best to interfere with these visitors in a friendly fashion.

—Colonel Gordon Cooper (NASA astronaut, tenth man in space)[1]

8.1 What Are They?

CARL SAGAN WROTE A FANTASTIC BOOK called *The Demon-Haunted World: Science as a Candle in the Dark* about the importance of critical thinking.[2] One part of this book that made a lasting impression on me was Sagan's "Dragon in My Garage Story." It goes something like this. Sagan claims to have a real fire-breathing dragon in his garage. So, he invites a rational-minded friend to come over and see the wonderous creature. Excited to see the dragon, the visitor waits with bated breath as the garage door opens. There's nothing there. Thinking the dragon might be smaller than expected or hiding, the friend looks carefully in the garage, but still nothing. "I don't see anything," the visitor finally says in dismay. Sagan retorts, "Oh, I forgot to mention that she's an invisible dragon." The rational visitor comes up with an idea, "Let's sprinkle a layer of flour on the garage floor so we can see the dragon's footprints." Sagan thinks that's an excellent idea but admits,

"Unfortunately, this particular dragon floats in the air and leaves no footprints." Becoming frustrated, the visitor proposes another more high-tech idea, "Let's use an infrared camera to see the dragon's invisible fire." Sagan counters with, "Yes, but this dragon's fire is heatless." And on it goes. Every way of physically measuring a characteristic of the dragon is met with a reason why that test will fail. The story concludes with a question:

> Now what's the difference between an invisible, incorporeal, floating dragon who spits heatless fire and no dragon at all? If there's no way to disprove my contention, no conceivable experiment that would count against it, what does it mean to say that my dragon exists?[3]

This story summarizes the ethos of the first part of this book. A claim is made that UFOs exist in our skies. And so, we play the part of the skeptical rational-minded friend and try to test this claim. The first question we might ask is: *Can we see them?* This corresponds to our eyewitness testimony category of evidence. Quickly followed by: *Can our instruments see them?* This equates to our evaluation of single and multiple sensor data. To eliminate the possibility of some kind of optical illusion, we also ask: *Do they leave behind physical evidence?* This equates to the physical evidence category. Can we prove or disprove the contention that UFOs exist in our skies? If they do exist, what are they? Do any of their characteristics imply a nonhuman origin? We must carefully review all the evidence we have collected over the course of Part I of this book and try to reach a rational and balanced conclusion.

In Part I, we examined four specific UFO incidents, namely, Japan Airlines Flight 1628, the Brazilian fragments, the Lonnie Zamora incident, and the Aguadilla object. Each case was quantitatively evaluated using four categories of evidence. Based on this metric, the case of the Aguadilla object rated the highest with an overall score of 59 percent, closely followed by the Lonnie Zamora incident with a score of 56 percent. The Japan Airlines Flight 1628 incident scored 43 percent. In last place was the Brazilian fragments case with an overall rating of 35 percent.

In two of the case studies, namely the Japan Airlines and Aguadilla object cases, UFOs were tracked on radar. In both cases, the UFOs did not transmit an identifying signal that is standard practice for civilian, military, and commercial aircraft. Except for the remote possibility of simultaneous radar malfunctions, or extremely unusual atmospheric conditions, what this implies is that pulses of radar energy must have interacted with some kind of physical object in order for them to have bounced back to the detector. In the case of the Aguadilla object, the UFO was also captured by an infrared thermal imaging video, with a clear and distinct heat signature.

What Does It All Mean?

In the Lonnie Zamora case, an object that remains unidentified to this day, despite concerted efforts from the US government and the FBI, landed and left physical imprints in the New Mexican sand. In the Brazilian fragments case, an alleged UFO exploded and left behind metallic fragments that have very unusual chemical properties, especially given when and where they were found. If this were the hypothetical dragon in the garage, we would have already seen it visually and with our instruments, detected its heat signature, and possibly even seen its footprints.

Given this evidence, the only objective and balanced conclusion is that UFOs probably are real physical objects. Such a conclusion is by now nothing new. The pentagon admitted the same thing in its June 2021 document entitled "Preliminary Assessment: Unidentified Aerial Phenomena."[4] The real question is, *What are they?* We will now review the evidence regarding the recorded flight capabilities of UFOs and their possible operation under intelligent control to seek clues that may help answer this question.

The UFOs in the Japan Airlines case and the case of the Aguadilla object exhibit extraordinary flight capabilities, as evidenced by their radar returns. The data suggests that the UFO in the Japan Airlines case undergoes accelerations that would essentially liquify a human occupant, due to the extreme g-forces. The structural integrity of all known aircraft, either manned or unmanned, would fail under these extreme conditions, resulting in its complete disintegration. Moreover, in both cases, UFOs tracked on radar underwent supersonic speeds with no reported sonic boom. This is a very puzzling characteristic that is difficult, perhaps even impossible,[a] to achieve with current technology.[5] Other flight characteristics of the Aguadilla object are equally difficult to explain. For example, the Aguadilla object performed trans-medium travel at high speed, going from air to water, without any significant change in speed. The Aguadilla object also had no apparent means of propulsion or lift, and even appeared to divide into two separate objects.

Based on this evidence, it is very tempting to conclude that these objects are not made by any government or private enterprise on Earth. However, I cannot conclusively exclude the possibility that these objects are highly advanced unmanned aircraft of some kind manufactured here on Earth. However, if this is the case, these advanced craft would represent a quantum leap in technology and material science, the magnitude of which is without precedent. It would also imply an unparalleled catastrophic failure in US national security that would surpass the intelligence failures surrounding 9/11 by several orders of magnitude.

---

a. There is at least one example of a cutting-edge aircraft, the X-59 QueSST, still in development by NASA and Lockheed Martin that can reduce the intensity of a sonic boom, but it cannot stop it all together.

We would be remiss if we did not consider another prosaic explanation, that the UFOs observed in our skies are due to a purely natural phenomenon, either known or unknown to science. One way of testing this hypothesis is to examine the evidence for signs of intelligent control. If an object appears to be intelligently controlled, it is unlikely to be a purely natural physical phenomenon. For example, you might encounter a river that flows between two almost perfectly straight banks for several kilometers before making a series of sharp thirty degree turns on its way toward a city. Such a river would imply its flow is under intelligent control since the river has a clear purpose and direction. Natural rivers are not so perfect, typically eroding material from the outside bend and depositing it on the inside, creating a rough curve with no sharp lines. Nature abhors straight lines. Another indication of intelligent control is mimicry. The ability of one object to observe another's behavior, understand how to change in order to match that of the other object, and then to precisely replicate the action, is a clear sign of intelligence. Let's review the evidence for the non-random deliberate movement of UFOs and whether they exhibit mimicry, which may help definitively decide whether they are a natural phenomenon.

In the Japan Airlines case, the UFO (see figure 2.6) moves along an almost perfectly straight trajectory for approximately thirty nautical miles. It then appears to perform a series of jumps back and forth, all of which are about the same distance. The UFO then confines itself within a smaller localized region for some time, tracing out an almost perfect semicircle during this period. This motion does not appear entirely random, and at times it seems deliberate and even precise. The movement of the Aguadilla object is more erratic, however, as shown in figure 5.1. Nevertheless, there is a concentrated cluster of movement that defines an overall trend toward the southwest. The infrared video of the Aguadilla object can also be used to recreate its flight path during the time it was recorded. The Scientific Coalition for UPA Studies[6] show that the most likely path is a clearly defined, and apparently deliberate, loop that passes through the Rafael Hernández Airport. Finally, going back to the Japan Airlines case, the pilot went on record saying that the UFO appeared to mimic the planes speed and direction exactly for a full seven minutes. Based on this evidence, I therefore doubt that the UFOs in these cases can be explained via natural phenomena alone, since they seem to display at least some degree of intelligent control. For a more direct example of UFOs displaying mimicry and intelligent control, see Commander David Fravor's account of his encounter with the by now famous "Tic-Tac" UFO during the 2004 *Nimitz* encounter.

So, let's try to summarize our understanding so far using a decision tree. A decision tree is a tool that helps visualize the decision-making process.

In this case, we want to arrive at a decision on what UFOs are. We do this by choosing between two mutually exclusive options. For example, we start with the assumption that there is a phenomenon, then we ask whether the phenomenon is man-made or not man-made. These options are mutually exclusive, since the phenomena can be either man-made or not, but it cannot be both. We then choose between these two branches on the decision tree and progress down to the next decision. A decision tree based on the evidence presented in Part I of this book can be seen in figure 8.1.

The three routes, I think, are most likely based on the evidence presented are shown by the thicker gray and black paths. Given that the enormous accelerations calculated in the Japan Airlines case would be fatal to any human occupant, I think it is much more likely that the UFO in this case was unmanned rather than manned. This eliminates two possible routes. Next, based on the evidence discussed regarding intelligent control and mimicry, I think it is more likely that the phenomenon is unnatural rather than natural. This eliminates two more paths. We are left with just three possibilities. Either the

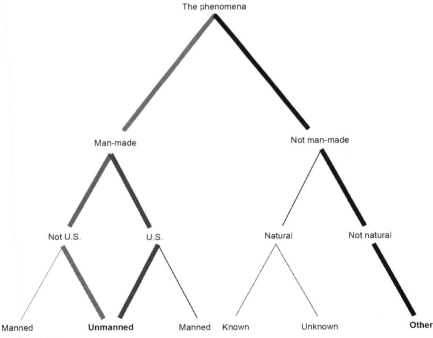

FIGURE 8.1
A decision tree for the phenomena. Based on the evidence presented in this work the most likely explanation for the phenomena is either unmanned non-US technology, unmanned US technology, or something other.

UFO phenomenon can be explained via unmanned US or non-US terrestrial technology or it falls into the *other* category.

Though I cannot rule it out, I find it unlikely that all UFO sightings can be explained as unmanned terrestrial technology. Here's why. At the time of writing, the United States is by far the most technologically advanced nation on Earth. And the US government has now categorically denied that UFOs are US technology.[7] In a strange turn of events, one would actually have to now believe a conspiracy theory, namely that the highest levels of the US government are lying to us, in order to maintain that UFOs are secret US tech. Radar data from as far back as the 1950s shows UFOs performing aerial maneuvers that even today's military aircraft are not capable of. For this technology to have remained secret and unused in combat for more than seventy years, plausibility becomes strained. Furthermore, for the US to test their most advanced technology over civilian populations, near commercial aircraft, and during military training exercises, seems both unlikely and reckless. Finally, the g-forces deduced from radar measurements of some of these UFOs are so high that no known aircraft could withstand them.

Therefore, I think the most likely explanation for the UFO phenomenon is a non-man-made unnatural *something*. That is as far as I'm prepared to go at this stage.

8.2 What Do They Want?

In Part II of the book, we cast a much wider net. The idea was to try to ascertain any possible motives behind the UFO phenomenon. Our approach involved investigating whether there was any connection between the number of UFO sightings per capita and five other often-discussed variables, namely, the military, water, the environment, earthquakes, and blood type. Our analysis revealed two possible correlations, and two definitive statistically significant correlations.

Two statistical tests were performed to quantify each possible association, the Pearson and Spearman tests. The first possible correlation discovered was between the number of UFO sightings per capita and the length of a country's coastline. Pearson's correlation coefficient implied a large positive correlation, with the p-value indicating the correlation was statistically significant at the 1 percent significance level. However, the Spearman correlation coefficient suggested only a moderate correlation, with the p-value showing it just misses out on being statistically significant. Crucially, the only possible correlation was with coastal seawater, no correlation was observed for freshwater. This result therefore remains ambiguous, but it is close enough to being significant to warrant further scrutiny. The second possible correlation uncovered was between the number of UFO sightings per capita and the number of earthquakes per square kilometer within the US. In this case, Pearson's test yielded a negligible statistically insignificant correlation. But Spearman's test found a statistically significant moderate positive correlation. Again, this result is inconclusive but deserves further attention.

The first clear and definitive correlation discovered was also perhaps the most unexpected. According to both the Pearson and Spearman tests, there is a large and statistically significant positive correlation between the percentage of a country's population with rhesus negative blood and the number of UFO reports per capita. The second definitive correlation is much less surprising. The number of movie releases featuring extraterrestrials against the number of UFO sightings reported worldwide per year also shows a large and statistically significant positive correlation in both statistical tests.

Leaving aside the two possible associations, what do the two more definitively established correlations tell us about the possible motives behind the UFO phenomenon? Well, perhaps nothing. Let's illustrate this point by thinking about ice cream and drowning. Yes, you read that right. Statistics show that whenever ice cream sales go up then the number of deaths by drowning goes up. Does ice cream cause drowning? Of course not. Then what is going on? It turns out that this is a classic example of a "lurking" variable. You see, the hidden, lurking, connection is temperature. As the temperature goes up, so do ice cream sales, but, crucially, so does the number of people that go swimming. Since more people are swimming, more people are likely to drown. Therefore, ice cream sales do not cause drowning. Instead, both ice cream sales and drowning increase in relation to a third "lurking" variable, temperature, which gives the illusion that one is causing the other.

Let's use this idea to help think about the observed correlation between the number of movies released featuring extraterrestrials against the number of UFO sightings. I would argue that there is a lurking variable in this case—namely, time. Figures 7.1a and 7.2a establish that both the number of UFO reports and the number of E.T. movie releases goes up with time, which gives the appearance of causation. There is an established correlation between these variables, however, there is not an established *causation*. What about the observed correlation between blood type and UFO reports? In this case, it is much harder to identify a lurking variable. Although one possibility that we have identified is mental health. Having Rh-negative blood has been broadly linked with mental health issues in the literature, and having a higher rate of mental health issues in each population may lead to more hallucinations of lights in the sky that are assumed to be UFOs. The connection between blood type and UFO reports, however, remains an interesting and open issue that warrants a focused effort to prove or disprove a causal relationship.

In Part II of this book, we also looked at the timing of UFO sightings. One interesting result from this analysis was a statistically significant spike in UFO reports on September 11, 2001. November 2019, the month often quoted in relation to the start of the coronavirus pandemic, was also found to have an uptick in UFO reports, however in this case it was not quite statistically significant. In another study, we found that UFO sightings seem to occur most frequently in June and July, regardless of whether we consider the northern or southern hemisphere. This implies it is not the increased number of daylight hours and time spent outdoors associated with the summer season that is responsible for more UFO sightings, but something else. We investigated one possible explanation in the form of an increased number of meteor showers in June and July, but the results were not convincing.

So, in conclusion, the motives behind the UFO phenomenon remain unclear. More work needs to be done to follow up some suggestive leads that were uncovered regarding the possible connection between UFO sightings and coastal water, earthquakes, and, most of all, rhesus-negative blood type. However, what I can tell you is that if you want to have the highest chance of seeing a UFO then the evidence suggests you should take as many rhesus-negative friends as you can find to a beach in Washington in June or July after watching a newly released movie featuring E.T.s.

8.3 What Now?

The field of UFO research is rapidly changing. Even while writing this book, four major steps forward have been taken. The first is that US senator for New York, Kirsten Gillibrand, has introduced an amendment to the 2022 National Defense Authorization Act (NDAA).[8] This amendment is revolutionary since it forces government transparency on the UFO issue. This act mandates the Pentagon create a new government body called the *Anomaly Surveillance and Resolution Office*, which is given the authority to pursue "any resource, capability, asset, or process" in the investigation of "unidentified aerial phenomena." The document also stipulates that

> Not later than December 31, 2022, and annually thereafter until December 31, 2026, the Secretary of Defense shall submit to the appropriate congressional committees a report on unidentified aerial phenomena. . . . The UAP office will be required to provide unclassified annual reports to Congress and classified semiannual briefings on intelligence analysis, reported incidents, health-related effects, the role of foreign governments, and nuclear security.[6]

The bipartisan amendment is cosponsored in the Senate by Senators Rubio (R-FL), Graham (R-SC), Heinrich (D-NM), and Blunt (R-MO).

The second significant advance came in the form of a 1,500-page database of declassified US government reports.[9] These reports were made public through a Freedom of Information Act (FOIA) request that was granted on April 5, 2022. It is clearly stated in the documents that "the evidence discussed includes scientific material that has been peer-reviewed." These truly fascinating and disturbing reports were put together by the Advanced Aerospace Threat Identification Program (AATIP), the secret US Department of Defense program that was officially terminated in 2012. A particularly interesting document from this database is entitled "Anomalous Acute and Subacute Field Effects on Human and Biological Tissues," which focuses on reported injuries to "human observers by anomalous advanced aerospace systems." The document lists the most important and well documented physiological effects such as "heating and burn injuries," "neurological effects," and "auditory and cranial nerve effects." The document rather unambiguously concludes that

> Sufficient incidents/accidents have been accurately reported, and medical data acquired, as to support a hypothesis that some advanced systems are already deployed, and opaque to full US understanding. . . . The medical analyses while not requiring the invention of an alternative biophysics, do indicate the use of (to us) unconventional and advanced systems.[10]

This is a powerful and shocking statement from the US government itself.

Thirdly, private scientific collaborations have been initiated. The Galileo Project was launched in July 2021 and was co-founded by the Harvard theoretical astrophysicist, Professor Avi Loeb. At the time of its launch, it had almost $2 million dollars of privately acquired funding, which is being put to good use. Project members, which now number more than one hundred professional scientists and computer scientists, are constructing the project's first multi-platform detector on the roof of the Harvard college observatory. The instrument will continuously scan the skies using infrared cameras, a radio sensor, an audio sensor, and a magnetometer. The aim is to use artificial intelligence to filter and scan the huge data collection, searching for that one smoking-gun image. Another recent private venture is known as UAPx. Founded by Kevin Day and Gary Voorhis, who were both serving military personnel aboard the USS Princeton during the 2004 Nimitz Encounter, UAPx now consists of a "global network of researchers, physicists, scientists, . . . and individuals dedicated to the scientific method in studying Unknown Aerial Phenomena." This organization has already made several interesting discoveries and hopefully will continue to do so.

Finally, on Tuesday May 17, 2022, the United States House of Representatives held the first public congressional hearing on UFOs for more than half a century. A historic event. Chairman André Carson, a Democrat of Indiana, opened the proceedings by stating that "unidentified aerial phenomena (UAPs) are unexplained, it's true, but they are real. They need to be investigated and many threats they pose need to be mitigated." During the hearing a video of an unidentified object was shown from the window of a US Navy fighter jet. The video, captured in 2021, showed an approximately spherical metallic-looking object that quickly passes by the cockpit of the aircraft. The deputy director of naval intelligence, Scott Bray, admitted "I do not have an explanation for what this specific object is."

The US governments involvement in, and understanding of, the UFO phenomenon seems to be being gradually disclosed. But we should not just sit and wait for the next government admission. We must collectively work on de-stigmatizing the UFO issue so that scientists can openly investigate the phenomenon without fearing for their job or fearing ridicule. I encourage each one of you to metaphorically take up arms. We must write to local authorities demanding answers, petition governments for further transparency, and offer our services to private efforts to help collect and analyze data. We must remain passionately curious, but we must also remain objective, and evidence driven.

One of the biggest questions we can ask ourselves is *are we alone*? Mankind has wondered about this question ever since we could wonder. It feels like we are closer than ever before to getting answers. Don't you want to be part of this effort? You just might become part of history. You just might spark a paradigm shift from a single *anomaly*.

> I think it's time to open the books on questions that have remained in the dark on the question of government investigations of UFOs. It's time to find out what the truth really is that's out there. We ought to do it because it's right. We ought to do it because the American people, quite frankly, can handle the truth. And we ought to do it because it's the law.
>
> —John Podesta (twentieth White House chief of staff)[11]

8.4 Notes

1. Gordon Copper, "United Nations Address" (New York, 1985).
2. Carl Sagan, *The Demon-Haunted World: Science as a Candle in the Dark* (New York: Random House). (1997).
3. Sagan, *The Demon-Haunted World*.

4. Office of the Director of National Intelligence (ODNI), "Preliminary Assessment: Unidentified Aerial Phenomena," June 25, 2021 (accessed October 10, 2021, https://www.dni.gov/files/ODNI/documents/assessments/Prelimary-Assessment-UAP-20210625.pdf).

5. Dr. Andrew May, "X-59 QueSST: The Quiet Supersonic Aeroplane that Could Revolutionise Air Travel," *Science Focus*, November 11, 2021 (accessed May 2022, https://www.sciencefocus.com/future-technology/x-59-quesst/).

6. Scientific Coalition for UPA Studies (SCU), "2013 Aguadilla Puerto Rico UAP Inident: A Detailed Analysis," *Scientific Coalition for UAP Studies*, last updated November 2, 2021 (accessed May 2022, https://www.explorescu.org/post/2013-aguadilla-puerto-rico-uap-incident-report-a-detailed-analysis).

7. ODNI, "Preliminary Assessment."

8. Congress, *S.1605 - National Defense Authorization Act for Fiscal Year 2022*, latest action December 27, 2021 (accessed May 2022, https://www.congress.gov/bill/117th-congress/senate-bill/1605/text).

9. Defense Intelligence (DI), "Anomalous Acute and Acute and Subacute Field Effects on Human Biological Tissues," *theblackvault*, March 11, 2010 (accessed May 2022, https://documents2.theblackvault.com/documents/dia/AAWSAP-DIRDs/DIRD_26-DIRD_Anomalous_Acute_and_Subacute_Field_Effects_on_Human_Biological_Tissues.pdf).

10. Ibid.

11. John Podesta, "'John Prodesta' Speech on Disclosing UFO Information at the National Press Club in Washington, DC" (Washington, DC, 2002). https://www.youtube.com/watch?v=ZU5JXguFYe8

Index

Page references for figures are *italicized*. An *n* following a page number refers to the note section.

9/11 and UFO reports, 140–141
abduction phenomenon, 137–138
Aerial Phenomena Research Organization (APRO), 40–42
Afanasyev, Victor, 89
Aguadilla object case, 75: flight path, *77*, 156; weather conditions, 81, 85n8; astronomical conditions, 82; anomalous behavior, 77–82
air traffic control (ATC): Japan Airlines case, 18, 21–24, 25, 32–33; Aguadilla object case, 75, 82
Anomalous Acute and Subacute Field Effects on Human and Biological Tissues, 161
anomaly: as a book, 6; definition of, 2–3; examples of, 1–2
Ariel school, Zimbabwe, 115

Byrnes, Special Agent, 59–60, 62, 68

Captain Terauchi, Kenju: career, 17, 24, 31; incident description, 17–22,
32; inconsistencies, 19, 20; language issues, 19, 31
closed time–like curves (CTCs), 138
Condon committee, 42, 44–45, 54n14
congressional hearing on UFOs, 162
Cooper, Gordon, 153
Coronavirus and UFO reports, 141–142
correlation is not causation, 126, 159
correlation: definition of, 90; classification by strength, *91*; Pearson's coefficient, 90–91, 95–97; Spearman's coefficient 97–98; r-values, 126–127; p-values, 94–95, 127; examples of, 91–94

decision tree, 156–157
DHC-8 Turboprop aircraft, 76, 78, 82
Dow, chemical company, 44–46, *52*
dragon in my garage, 153–155

earthquake lights (EQL), 120–121
earthquakes and hallucinations, 121

Elizondo, Luis: AATIP involvement, 3–5; New York Times article, 3

Federal Aviation Administration (FAA), 17–18, 25, 31–33
FLIR (camera), 4, 10
Fontes, Dr Olavo, 40–41, 43
foo fighters, 102
Fravor, David, xii, 6, 10, 156
French UFO study, 116

Galileo project, 161
g-force, 29, *30*, 32, 155, 158
Gillibrand, Kirsten, 160

Heaversine formula, 78, 81, 84
Hellyer, Paul, 17, 37n1
Hemolytic Disease of the Newborn (HDN), 124
Hoover, J. Edgar, 135
Hughes aircraft, 62–64
Hynek, Allen, 9, 13n16

intelligent control (of UFOs), 155–156, 157
International atomic energy agency (IAEA), 117
isotopes, 47: definition of, 47; of magnesium, 47–49; of rubidium, 52; sigma tension, *48;* terrestrial ratios, 47, *48*, 53

Jacobs, Robert, 103–104
Japan Airlines incident: alternative explanations, 31–33; key events, *23*; radar detection, 19, 20–21, 23, 24, 26, 32–33; radio interference, 19

Kean, Leslie, 3, 13n6, 54n1
Kuhn, Thomas, 2, 12n2

Lorenzen, Coral, 41–42, 54nn6,7,10
lurking variables, 159

magnesium: aeronautical uses of, 46; properties of, 46; *See also* isotopes
Masters, Michael P., 137–139
McDivitt, James, 75
Mitchell, Edgar, 57, 73n1

NASA, 62, 108
National Defense Authorization Act (NDAA), 160
National UFO Reporting Center (NUFORC), 99
Nimitz encounter: description of events, 4, 104; evidence, 10–11, 156; impact, xii, 156, 161
North American Aerospace Defense Command (NORAD), 19, 21, 33
Novikov self-consistency principle, 138
nuclear weapons (and UFOs), 103, 108, 132n12, 144

Obama, Barack, 1, 12n1

paedomorphosis, 136
Pentagon: June 2021 report, 5, 155; UFO videos, 4–5
physical evidence: Brazilian fragments case, 42, 49, 50–52; definition of, 11; Lonnie Zamora case, 61, 65, 67–70

QJQ radar site (Pico Del Este), 76–78, 83

radar: specifications, 25, 76–78; split radar effect, 32–33; transponder, 21, 26, 32, 75, 83; types of return, 21, 25–26, 32
Rafael Hernández Airport (BQN), 75–76, *77*
Regional Operations Command Center (ROCC), 19–21, 23
Reid, Harry, 5, 39, 54n1

Index

Salas, Robert, 103
Scientific Coalition for UFOlogy (SCU), 75, 82–83, 85n3
Shag harbour incident, 109–110
SJU radar site (San Juan), 83
sonic boom, 27, 29, 32, 78, 83, 155
Spent nuclear fuel (SNF), 116–117
standard deviation, 48, 53, 140–142
student hoax (hypothesis), 65–66
Sued, Ibrahim, 39–41
surveyor module (hypothesis), 62–63, 64, 65

thermal imaging video, 76, 78, *79–80*
time travel hypothesis, 137–139

UAPx, 161
Ubatuba fragments: chain of custody, 40–42; description of, 41; explosion evidence, 49–50; impurities, 45, 50–51; isotopic content. *See* isotopes. magnesium content, 42–45, *44;* strontium content, 45–46

UFO reports by US state, 99, 100, *101*
Unidentified Aerial Phenomena Task Force (UAPTF), 4
US Customs and Border Protection, 76, 78, 85n7

Walton, Travis: abduction, 135–136, 151n3; description of beings, 136; physical evidence of incident, 136
wartime UFO sightings, 102, 144
Weight, from indentation: alternative explanations, 67–68, 69; assumptions, 70–71; mass estimates, 62, 67; method, 70–72
worldwide UFO reports: from 1940–2021, 142, *143*; exponential fit, *143*, 150; logarithmic scale, *143*

Zamora, Lonnie, 57: career, 57, 61, 65; description of incident, 57–59; life after incident, 61; reliability of, 68

About the Author

Daniel Coumbe received a PhD in theoretical particle physics from the University of Glasgow. He has held research positions at Syracuse University in New York, Jagiellonian University in Poland, and the prestigious Niels Bohr Institute in Denmark. Coumbe has published fourteen peer-reviewed research papers on theoretical physics, including articles in world-leading journals such as *Physical Review Letters* and *Classical and Quantum Gravity*. Dr. Coumbe is the author of a graduate-level textbook on quantum gravity, *Magnifying Spacetime: How Physics Changes with Scale*. He has taught college-level courses in physics and mathematics and has given numerous presentations at international physics conferences.

Coumbe was born in England but currently lives in Denmark with his wife and two children.